# Cultivating Commons

# Cultivating Commons

*Joint Ownership of Arable Land in Early Modern Japan*

**Philip C. Brown**

University of Hawai'i Press
Honolulu

**Library of Congress Cataloging-in-Publication Data**
Brown, Philip C., 1947-
Cultivating commons : joint ownership of arable land in early modern
Japan / Philip C. Brown.
p. cm.
Includes bibliographical references and index.
ISBN 978-0-8248-3392-3 (cloth : alk. paper)
1. Commons—Japan—Niigata-ken—History.   2. Farm tenancy—
Japan—Niigata-ken—History.   I. Title.
HD1289.J3B76 2011
333.2—dc22
2010037625

Designed by Wanda China
Printed by The Maple-Vail Book Manufacturing Group

*In memory of Matsunaga Yasuo, diligent scholar, generous friend*

# Contents

Color plates follow page 144

# List of Tables

# *Preface*

This project has dual origins. In the long run, it comes from a long-standing interest in rural society. Like many an urban and suburban child, I had little contact with rural society growing up: that is, until my father, looking for a new challenge in life, slid into farming as a major pastime. While I never possessed my father's physical energy (between working at Eastman Kodak and the farm in the Finger Lakes region of upstate New York, he routinely put in twelve- to fourteen-hour days, exclusive of commuting time, four to six days a week), working with him I came to know farmers: people who labored in the face of natural challenges, put in long days, and became their own large animal veterinarians, mechanics, and soil scientists. I gained a new appreciation for the value of informal as well as formal education and respect for the intelligence and diligence that many farmers brought to their work. From that time on, I was aware that farming and rural life provide many stories worth telling, although I had no particular plans to be their raconteur at the time.

My first book, *Central Authority and Local Autonomy in the Formation of Early Modern Japan: The Case of Kaga Domain*, explored rural matters and provided a chance to act on my youthful observations. My initial interest lay in the early modern foundations of Japan's late-nineteenth- and early-twentieth-century economic transformations. The choice of Kaga domain as a case study proved fateful in two ways.

First, my research experience there transformed my initial dissertation project into one that focused on a much earlier period and the process of political stabilization in the late sixteenth and seventeenth centuries.

The second serendipitous outcome of my Kanazawa sojourn and the second stimulus for this project was exposure to *warichi*, a system of joint village landholding and redistribution. At the time, I thought it might make an interesting article at a future date. While I incorporated some discussion of it in my first book, I came to understand that joint ownership was much more widespread than Kaga, and, encouraged by Margaret McKean, John Richards, and others, I extended my investigations, explored *warichi*'s historical and theoretical implications, and ultimately expanded my project into this book.[1]

Although the circumstances of farming in early modern Japan (ca. late sixteenth to mid-nineteenth century) were considerably different from those in twentieth-century upstate New York, the study of *warichi* promised insights into rural society and a chance to tell something of the story I encountered as a teenager. Rice paddy agriculture in a pre-modern age lacked many elements of modern agricultural regimes, but success in both requires hard work, thought, careful planning, and the knowledge that in the final analysis, key requirements for success are outside one's control: rainfall, temperature, and natural calamities like severe hail, floods, or landslides.

Joint landownership has long been thought of as one means of dealing with nature's uncertainties—a proposition I will take up below. As a result of my interest in testing the degree to which topography and climatic conditions interacted with *warichi*, I found myself drawn to the use of a relatively new technology, geographic information systems (GIS), and the satellite-based technology of the global positioning system (GPS). These tools allowed me to make many of the maps that appear in the following chapters. Absent these two technologies I would not have had any source other than anecdotal evidence with which to explore the relationship between villages, their natural surroundings, and *warichi*.

Understanding joint landholding practices leads to insights broader than those associated with the relationship of a village to its environment, however. These include greater understanding of villager attitudes toward the land and the reach of state power, whether situated in a national or domain context. It is to these realms as well as landholding, a central element in the lives of villagers, the most populous class in early modern Japan, that I turn below.

# A Note on Style

*Japanese names* appear in the customary Japanese order, family name first, personal name second. During the period prior to the Meiji Restoration (1868), commoners did not use surnames.

*Romanization* of Japanese is straightforward, with the following conventions followed for vowels: "a" is always pronounced as the "a" in "about," "i" is always pronounced as the "ea" in "each," "u" is pronounced like the "o" in "whose," "e" is always pronounced as "e" in "excellent," and "o" is always a long "o" as in "open." The vowels "u" and "ū," "o" and "ō" are different; the vowel with the macron takes twice as long to say as its partner without a macron.

*Dates* are presented as year.lunar month.day in which years have been converted to the current Western calendar, but the month and day of the lunar calendar have not, so 1861.3.1 is the first day of the third lunar month of 1861 (Western calendar). Japanese era names are sometimes used in place of the Western year, giving the era name.lunar month.day, thus Genna 6.4.3 is the third day of the fourth lunar month in the sixth year of the Genna era.

*Weights and measures* were not standardized for all of the periods addressed below, and it is therefore not possible to present precise modern equivalents. However, the conversions below represent a fair approximation.

### LINEAR

| | |
|---|---|
| 1 *sun* (a Japanese "inch") | |
| 10 *sun* | = 1 *shaku* (a Japanese "foot") |
| 6 *shaku* 3 *sun* | = 1 *ken* (about 2.2 yards, 1.8 meters) |

### AREA

| | |
|---|---|
| 1 sq. *ken* | = 1 *bu* (about 4 sq. yards; 3.5 sq. meters) |
| 10 *bu* | = 1 *se* |
| 10 *se* | = 1 *tan* |
| 10 *tan* | = 1 *chō* (about 1 hectare or 2.5 acres) |

### VOLUME

| | |
|---|---|
| 1 *koku* | = 5 English bushels; 180 liters |
| 1 *to* | = .1 *koku* |
| 1 *shō* | = .01 *koku* |
| 1 *gō* | = .001 *koku* |

# Acknowledgments

As all authors of research monographs know, extended projects require the cooperation and support of many people and organizations. In repeated visits to Niigata for research on this book, independent local historian Matsunaga Yasuo frequently took me under his wing and introduced me to scholars, local history editorial offices, and local archives throughout the central and southern Echigo region. To be sure I had a gut-level feel for Tokugawa and early-twentieth-century rural life, he fed me only simple fare when we dined at his home, each dish accompanied by explanation and some insight into rural life. I looked forward to presenting him a copy of this book, a pleasure stolen by his sudden illness and death in February 2010.

Niigata University has served as my base for multiple research trips, and Kobayashi Shōji (Niigata and Teikyo universities) has been a warm and welcoming host. His invitations to conferences and introductions to local historians provided me with new perspectives and stimulation. Yoshii Ken'ichi, also of Niigata University, patiently ferried me back and forth to the Nagaoka City History offices on his periodic research trips during my first extended stay in Niigata and has continued to support my research on this project.

Fukuda Chizuru, Andō Masahito, Yamada Tetsuyoshi, Aoki Mutsumi, and Watanabe Kōichi hosted me at National Institute of Japanese Literature, Division of Historical Manuscripts, during late 1997 and early 1998 under the auspices of a Ministry of Science, Education, and Culture Center for Excellence Grant. Professor Fukuda in particular offered support as I puzzled through the manuscripts of the Satō Family Documents collection.

Especially in the early stages of my research, Aono Shunsui, formerly of Fukuyama, Hiroshima, and Ehime universities, was particularly encouraging.

Key elements of my argument have benefited from the use of geographic information systems (GIS) technology, a venture that has led to exciting interdisciplinary contacts with historical geographers, historians in a variety of regional-national fields, and even engineers. While I am very grateful for the assistance of many individuals, I particularly want to acknowledge Duane Marble, professor emeritus of the Ohio State University Geography Department, who introduced me to GIS. In Japan, Professor Yano Keiji along with his colleagues and graduate students in the Geography Department of Ritsumeikan University have also assisted me with the use of GIS technology. One of the students, Kawasumi Tatsunori, who has now joined his mentors on the faculty, was notably helpful in the late stages of map preparation. Hiroshima University Geographer Emeritus Ishida Hiroshi and Professor Mizoguchi Tsunetoshi (Geography, Nagoya University) have also provided encouragement and support.

Two Duke University faculty provided extended support over the years. No one individual has been more influential in stimulating my research on *warichi* than Margaret McKean of the Department of Political Science. She has been a rigorous but supportive critic, generously sharing her expertise in common pool resources. The late John Richards (History) consistently included me in conferences on property rights, economic development, and the sustainability of natural resources. Both helped me appreciate broader contexts and theoretical issues to which the study of Japanese joint landownership contributes.

At The Ohio State University, Mansel Blackford has taken a consistent interest in my work as a parallel to some of his own research interests in the ecological history of the United States. He has kindly read the entire manuscript of this book and given me timely comments. Two others, James Bartholomew and Donna Guy, generously read the manuscript in part or in its entirety.

Several additional colleagues have offered helpful comment and advice. At quite early stages, Patricia Sippel (Toyo Eiwa Women's University) and John Van Sant (University of Alabama, Birmingham) made comments that helped me restructure and direct my argument. Mary Elizabeth Berry (University of California, Berkeley), Karl Friday (University of Georgia), and Brett Walker (Montana State University) provided helpful observations on the penultimate draft of the manuscript.

I have received financial support from a variety of sources over the years that made repeated and extended fieldwork possible: The U.S. Department of Education Fulbright-Hays program, the Fulbright program administered by the Institute of International Education, the Japan Foundation, the Social Science Research Council, the Japan Society for the Promotion of Science—National Science Foundation Short-Term Fellowship Program, the National Science Foundation Program in the History and Philosophy of Science, Engineering, and Technology, the Seed Grant program of The Ohio State University, and the American Philosophical Society. In addition, the College of Arts and Humanities of The Ohio State University has provided support through its publication subvention and manuscript preparation support programs. I hope they will find the fruits of my work commensurate with their investments.

To all of these groups and individuals, and more unnamed, my heartfelt thanks.

Elements of this project have been published in a variety of different venues. However, pulling this early material together to develop the arguments below brought to my attention inconsistencies in past treatments: much of what I did not catch, my readers did. If, despite the efforts of my generous colleagues, errors of fact or interpretation remain, they are solely my responsibility.

# 1

## Introduction

Many early modern Japanese villages (ca. 1580–1868) failed to adopt anything like private individual landownership, yet they still experienced economic diversification and growth. In some 30 percent of Japan, no direct link existed between a farmer and the land associated with his cultivation rights. Especially in many parts of the Hokuriku, Shikoku, and Kyushu regions, the land one farmed at any given time was determined by village or baronial domain *(han)* policy employing one of three mechanisms: allocation of cultivation rights linked to family composition, a fixed order of rotation, or lottery. Although there are three major variants of these systems (per capita allocation, allocation per family, and per share allocation), and many lesser ones, I subsume all under the most widespread Japanese moniker *warichi*, literally "dividing the land," and under the term "joint ownership" in English. These forms of tenure applied to arable lands, not to the commons *(iriai-chi)*, which sometimes employed similar mechanisms to allocate access to a resource.[1] Depending on which variant one discusses, these tenurial regimes not only determined who farmed which plots, but also imposed severe restrictions on alienability of rights in land.

The presence of these joint ownership forms complicates a number of common images of early modern and modern (post-1868) Japanese society, economy, and government as well as historians' assessment of the balance between continuity and discontinuity during the two eras.

1

Early modern Japan evinced an array of property rights, many of them joint forms of ownership, not just private property in which the owner exclusively controlled all facets of land use including alienability. Joint owners often had no need to change property rights regimes in order to foster economic growth during the early modern era. Indeed, in some cases joint ownership persisted well into the post–World War II era, and even without the force of law or written constitution these forms of joint ownership functioned well. Despite the extensive use of joint village ownership and its implications for the structure of ownership rights, economic growth, and farmers' sense of attachment to the land, it has not received much attention in English-language studies.[2]

Japan's experience with joint landownership not only presents conundrums in its own right, it challenges some of the ways social scientists and others commonly think about property rights, their relationship to economic growth, natural resource management and preservation, and contemporary social policy related to these realms. We need to consider an array of property rights, not just two monolithic categories of "private" and "public." As discussed below, joint ownership of arable land fits in between these two poles.

In addition, Japan's history makes it clear that these forms of ownership are not mutually exclusive. They can peacefully, even beneficially, coexist in the same economy. Parts of early modern Japan witnessed some form of public ownership and fee simple ownership, but large portions of the land were under various forms of joint ownership of either arable lands, the commons (*iriai-chi*, meadowlands, forests, and some other resources), or both.

The structure of ownership rights was adapted to serve particular functions and was adaptable over time. While fee simple ownership structures served many agricultural communities, joint ownership was the universal option for resources such as the commons that were essential for all cultivators. Joint ownership was adapted to arable lands to serve other special functions such as spreading natural hazard risk or providing a special resource. Sometimes joint ownership of arable lands was ephemeral, but more commonly what stands out is its long-term persistence, even into Japan's modern era. Its longevity suggests another underappreciated characteristic of property rights—their malleability and ability to adapt successfully to changing circumstances.

Some of the complexities regarding property rights may have escaped the attention of Japanese historians, broader theoreticians, and policy-makers because most are products of urban, capitalist environments; while discussion of the modern—economic, social, and political—focuses heavily on urban-centered activities, in premodern societies and states, wealth was overwhelmingly agricultural. Developing effective means of controlling access to land posed a key challenge for ruling classes and the wealthy. The control of arable land provided a firm economic source of government revenues and a conservative, secure foundation for personal wealth. That wealth remained central for most of the world well into the twentieth century, a world far different from that of today's urbane thinkers and policy wonks. Control of landed wealth, however, was only one of the major challenges faced by rulers, owners, and cultivators of premodern arable lands.

For these societies, confronting variable natural circumstances comprised the most fundamental challenge to successful agriculture. The crucial event in the development of human societies must surely be that transition variously called the Neolithic or the first agricultural revolution during which humans learned to farm. Agriculture's ability to produce storable surplus calories became the cornerstone of stratified societies with diverse occupational structures and urban centers. Yet these benefits could be exploited only to the degree that agriculturalists developed relatively effective tactics for dealing with the vagaries of nature.

Today, those of us in the developed world presume the bounty of agriculture. Typically we are only marginally inconvenienced by higher prices at the grocery store when California lettuce or Florida oranges are damaged by foul weather. We have lost sight of the struggles of farmers in the ages when commercial fertilizers, labor-saving tractors and machinery, and a host of other modern agricultural supports were absent. We place great faith in engineered control of much natural risk—buildings, dikes, dams, and the like. While to a great degree the modern industrialized world has lost its sense of dependence on the whimsy of nature, the damage inflicted by disasters such as Hurricane Katrina and the Indian Ocean tsunami forces us to acknowledge nature's impact. Until such recent, large-scale ravages, our understanding of nature's power has often come from novels like Jack London's *Call of the Wild* or Ernest Hemingway's *The Old Man and the Sea* and movies like *The Perfect Storm* that provide vivid depictions of individual battles that represent the broader struggles of humans with nature. Images of hunters, fishermen, and whalers are misleading in one sense: they distract our attention from the dependence of human beings on natural forces throughout recorded history, especially in the essential albeit mundane world of the farmers who composed the overwhelming majority of human populations virtually everywhere until the most recent century.

Among natural forces affecting agriculture, those related to weather are most central, and of these, water supply and temperature are perhaps most critical for the production of a given crop. There is little that farmers can do about temperature, but water supply has long been amenable to amelioration through various strategies.

Water supply in agriculture is a Goldilocks problem: crops do poorly with either too much or too little water, or if appropriate amounts come too early or too late in the growing season. Any given crop does best with a "just right" amount of water at "just the right" time. Monsoon societies, East Asian societies and Japan among them, contend with a broad range of precipitation variability and its associated over- and under-supply of water, a range broader than is typical in Western agriculture. They have long engaged in a wide array of riparian engineering and irrigation construction to ameliorate these swings. To deal with the challenges of too much water (flooding and frequently, if indirectly, landslides) people built dikes and dams.[3] To provide water in a timely fashion, societies learned to store water and irrigate, even learning to lift water uphill in a variety of ways.

In addition to employing technological solutions to floods and water supply problems, many societies learned to ameliorate the challenges of nature through laws and regulations—social engineering. Indeed, throughout much of history the limitations of civil engineering technology encouraged social rather than technological solutions. Among them, zoning and restricted construction in hazardous areas constitutes one practice that has deep historical roots and survives today.

Agricultural communities in Europe as well as Asia developed practices that effectively created a diverse portfolio of lands for each cultivating family. Each family consequently farmed widely scattered fields that were subject to different natural risks.[4] As a result, the fate of a family's annual sustenance did not depend on the output of only one field or one type of field. In addition, in a largely self-sufficient family economy, farmers avoided monoculture, instead cultivating a wide variety of crops—oats, wheat, cucumbers, squashes, eggplant, soybeans, to name just a few—so that their food supply was not dependent on successfully growing a single crop.

When technologies, avoidance of monoculture compelled by self-sufficiency, and informal acquisition of lands scattered throughout a village proved inadequate, some societies developed other socially engineered structures to control exposure to natural hazards or to ameliorate their impacts. Among such practices, both administrators writing in the Tokugawa period and historians writing thereafter have viewed the development of *warichi* as a means by which interdependent village populations shared in the damage from floods, landslides, and changes in soil fertility. Regardless of whether or not this was the intent of joint ownership, the most common forms of *warichi* spread natural hazard risks and the benefits associated with microclimatic variation in a village among all shareholders. They prevented concentration of the best lands in the hands of a few families and created a diverse portfolio of lands of all types for each shareholding family. The well-known Russian *mir* offers a parallel instance in a different cultural and agricultural context.[5]

To the degree that joint ownership of land and other resources is linked to debates over how to manage a natural resource effectively, deal with natural hazard risk, and adapt to transformations of the natural environment, study of Japan's experience reminds us that nonmarket and nontechnological social solutions have been viable and may still be viable today. As we adapt to global warming—whatever its causes—and live with the outcomes of major engineering projects designed, for example, to regulate water supply or adapt to changing weather patterns, we have become aware that while technology may help use resources more efficiently, promote a healthy environment, and make other positive contributions, it is not a panacea. Risks remain and cannot be engineered out of existence. Indeed, engineering solutions often have unanticipated risks associated with them. We know, for example, that extensive diking of the Mississippi has contributed to recent major flooding. While early warning systems can help us predict the impact of a tsunami, they are not a complete solution. Exploration of long-term success with social solutions such as Japan's joint ownership of arable land increases our awareness that well-designed social institutions can make useful contributions to solving contemporary policy issues; they can provide a measure of security and equity for participants in dealing with natural hazard risk.

If we are to explore humanity's experience more fully and understand a broader array of property rights and resource management techniques, we need to pay attention to past agricultural societies; yet the challenges farmers faced and the process of farming hold limited attraction for the vast majority of twenty-first-century historians. There have been periods in the past when historians showered great attention on agriculture (the work of M. M. Postan, among others, is a particularly noteworthy example, along with that of Esther Boserup), and while there are still significant, active communities of scholars who explore North American and European rural history (often with a focus on the environment), a bibliographic survey of English-language studies on Japanese history reveals a rather short list of both publications and authors who research agricultural activities and communities. The most noted of these would be Thomas C. Smith.[6]

Within the study of past agricultural societies, exploration of the relationship between property rights and economic development has been a major interest. For example, among economists, Nobel laureate Joseph Stiglitz has explored the conditions under which share-cropping might provide better incentives than payment of rents in cash and vice versa.[7] Gary Libecap, among others, has explored why property institutions that seem ill suited to maximizing economic utility survive. Directing his attention specifically toward property rights for natural resources, he explores the considerable variety of arrangements that arise even within similar legal contexts.[8]

In addition, in the context of the broad historical movement toward modern economic development, government leaders, economists, and economic historians often laud the value of private property's economic incentives to maintain and improve arable land. From the English enclosure movements that ultimately led to the late-eighteenth-century Enclosure Acts down to the present day, joint ownership or use of arable land by many people has often been viewed as largely wasteful in burgeoning market-oriented economies. Pressures to permit more efficient adaptation to market conditions by restricting decision-making rights in land to an individual or a few owners underlay disputes over land in a wide variety of geographic and historical contexts. Local movements in this direction in England predated the formal parliamentary acts of enclosure. Here the effort was to limit the role of many tenant users on land that was already privately owned.[9] Similarly, disputes over the efficiency of the communitarian *mir* were ultimately brought to a head in the Stolypin Reforms in late imperial Russia.[10] Conflicts between beneficiaries of communal land rights and those seeking to end them have been prominent in Mexican history.[11] For modern command economies, too, state ownership that divides outputs among many people based on needs (socialism) has been regarded as clearly retarding agricultural outputs in the Soviet Union, the People's

Republic of China, and other communist regimes.[12] In between state ownership and private, fee simple ownership lies the mechanism some Chinese regions chose to move toward private individual ownership of land: a system of long-term leases that in some ways mirror the functioning of the *warichi* system I explore here, an arrangement that gave individuals fuller control over the land they leased than under the old commune structure.[13]

Each of the preceding cases evinces a tension between control of land use by many and control by an individual or small group, but they also demonstrate the need for more subtle distinctions among rights in land than just a bifurcated conception of state or community ownership versus private ownership. Societies have been very creative in generating a wide combination of public and private rights in land; simple condemnation of control by many as inefficient and praise of control by an individual as efficient does not do this variety justice. Our scholarship should acknowledge these variations.

The negative critique of common lands has been formalized in Garrett Hardin's conception of "the tragedy of the commons."[14] Hardin, a biologist concerned with the effects of overpopulation in a context of finite resources, explored ways to maximize long-term viability of natural resources. He argued that there was no technological solution to resolving the conflict between resources available and population pressures, and that policy choices would have to be made regarding what society maximized since even the idea of maximizing benefits for the largest number of people would fail. In exploring options for decision making about the consumption of resources, he sought to test the idea that the public good would be provided for if individuals were left to decide what was best for themselves as individuals. He explored the incentives for individuals to exploit or maintain what he called the commons. He argued that individual actions on common property led inevitably to exhaustion of the resource (he used free grazing lands as his key example) because short-term incentives and disincentives were structured to emphasize near-term gain rather than long-term maximization of resource use. He concluded that unrestrained private action on the commons leads ultimately to the destruction of the resource. That is, in the absence of private ownership, in any commons, only force, externally imposed, could prevent exhaustion of a commons resource (an inherently expensive and therefore economically inefficient proposition).[15]

This is the classic "free rider" problem, in which one actor makes maximum use of a resource to the detriment of other users without paying any cost or penalty to produce or maintain a public good. In one formulation, "individual consumers will fail to state publicly their full monetary evaluation of a collective good."[16] It presumes that people seek to get something for nothing from a public good.

Hardin's argument that, in comparison to the commons, private property provides superior incentives to preserve a natural resource has a mirror image in scholarly arguments that private, individual property ownership (fully exclusive, fee simple) provides maximum incentives to invest in improvements. Other scholars have also taken up a wide set of issues surrounding the establishment of property rights and the interaction between economics and law. Nonetheless, these analyses largely reinforce the well-established belief in the economic efficacy of private property rights.[17]

Critiques of Hardin have been vigorous, and a large number of them note the diversity of arrangements under which successful management of common pool resources has taken place. While counterexamples have been developed, the chief complaint about his model lies in its assumption that "the commons" is not regulated by those who exploit it. His model of grazing/overgrazing is based on a Wild West view of the movement of cattle barons on the Great Plains of the American West rather than exploration of the long-standing examples of internally regulated commons found throughout world history.[18]

Joint ownership, represented by the commons and its purported propensity for free rider problems, has been contrasted with the operation of the modern corporation, an organization that also involves management by many but is presumed to eliminate free rider problems through properly structured incentives for participants. Yet this presumed difference has also come under attack—from social scientists, not from humanist historians familiar with long-standing traditional institutions. For example, Gary Miller, an organization specialist, employs game theory and other approaches to argue that free rider problems are inescapable in corporate hierarchies; the opportunities to free ride are many, and there is no incentive system that can eliminate them. Nonetheless, the advantages to corporate cooperation overcome the costs of carrying free riders, as the profits of companies like Exxon show. Consequently, Miller goes on to argue, the only way to further the best interests of the corporation is through leadership that successfully gets employees to sacrifice short-term benefits for the corporate interest.[19] Such a conclusion suggests that perhaps the contrast between traditional joint management and modern corporations is not as great as has been presumed, and therefore traditional ownership forms like *warichi* might actually be economically efficient, despite any costs of free riding.

Japan's nineteenth-century political transformation, the increase in state power, and economic diversification and growth constitute major themes in the postwar study of Japanese history, and interpretive frameworks similar to those outlined above play a prominent role. Whether viewed positively as a foundation for Japan's postwar revival and economic success, negatively as the basis of prewar imperialism and political oppression, or in some combination, these themes and their dynamics occupied the attention of postwar historians of modern Japan and encouraged exploration of their Tokugawa roots. Indeed, the conviction that the Tokugawa period held the keys to explaining Japan's late-nineteenth-century capacity for self-transformation encouraged its recharacterization as Japan's early modern era.[20]

The search for the roots of Japan's modern economic transformation led scholars to emphasize pre-Meiji activities and organizational patterns deemed compatible with the development of modern economic and business organization. For example, Thomas Smith stressed the increased market orientation of farm enterprises, a theme elaborated by Edward Pratt.[21] Demographic studies suggested that transformations of the family—increased presence of a nuclear or stem family structure, the declining prominence of large family patterns, and the consequent freeing of dependent labor—maintained and improved standards of living based on the development of a more flexible family economic unit. The decline of long-term contract labor arrangements in favor of day labor also has been cited as a movement toward modern business and economic flexibility.[22]

Economic historians tell a largely triumphalist tale of the emergence of private property and capitalism in Japan that parallels early modern Europe's trajectory.[23] Put schematically, in Japan's case, the story begins in the late sixteenth and early seventeenth centuries, grounded in castle town growth, the emergence of a national market, and, in rural areas, the development of commercialized agriculture during the long Pax Tokugawa (ca. 1600–1868). Such developments, seen as forward-looking and providing a secure foundation for the full-scale emergence of a market-oriented economy, transformed late-nineteenth-century Japan into a world-class economic and military power and ultimately led to the Pacific Wars of the 1930s and 1940s.

For example, Hayami Akira drew on Robert Heilbroner's concept of an "economic society," a "market economy…in which the providers of finance, goods, and services and their consumers, act to the greatest degree on economic values," to describe the nature of the economic transformation that enveloped Japan during the Tokugawa era.[24] In other words, economic actors as a group behaved in an economically rational, profit-maximizing manner that generated new techniques of production and wealth, especially in the nineteenth century. In agriculture, this orientation led to what Hayami called an "industrious revolution" of labor intensification made worthwhile by the prospect of farmers garnering the benefits of resultant increases in output. A similar emphasis characterized studies from the Quantitative Economic History Group.[25] Saitō Osamu showed how farm families took in and sent out labor, structured inheritance, and married in patterns that adapted to changing economic circumstances to maximize survival and prosperity. Takagi Masao elaborated the relationship between family structure and famine survival in a similar manner.[26]

Although these scholars are arguing that farm families increasingly participated in market-related activities and did so in more economically rational ways, they are not describing the birth of farming enterprises fully dependent on the market, a capitalist farm management. While farmers produced large quantities of potentially commercial crops—rice and soybeans in particular—they were not free to sell much of this output directly on the market. Most rice, and in some parts of the country much of the soybean crop, was used to pay the land tax, villages' major financial obligation. As the Tokugawa era progressed and per hectare yields rose, more and more rice, soybeans, and other crops were available for sale on the market, but until the Meiji Land Tax Reforms of the 1870s, cash payment of the land tax was restricted. This system of taxation limited a farm enterprise's flexibility in exploiting a region's natural economic advantages and adapting to an increasingly integrated national market. With the Meiji Land Tax Reforms, Japan adopted a British model of assessing fixed taxes payable in cash and based on the farmland's market value, a situation that thrust all farmers fully into the market economy.

Studies of early modern Japan have also reflected a positive evaluation of private property rights and associated them with a modern society. They link Hideyoshi's late-sixteenth-century and similar surveys to the creation of villagers' full, exclusive rights to land, akin to exclusive private individual ownership. Except for limited exercise of

rights similar to eminent domain, rights in land could only be transferred among farmers as a rule.[27] For example, noted legal historian
Ishii Shirō argued that although Hideyoshi's surveys were not detailed
enough to provide a basis for "title" to the land, with the passage of
time and the development of greater accuracy in survey procedures—
repeated resurveys that exposed unregistered lands, corrected errors,
and so forth—the surveys did provide a basis for defining a peasant's
claim to a piece of land.[28] Kozo Yamamura, an American economic historian, argued that while a variety of considerations moved parts of six-
teenth-century Japan toward private landholding arrangements, there
was no systematic mechanism to advance this policy nationally until
Hideyoshi's land surveys of the late sixteenth century. "In a legal and
political sense, the *honbyakushō* system [reliance on 'basic peasants' and
the village organizations they controlled] made newly 'listed' taxpay-
ers of many of the 'dependent agricultural laborers,' marginal tenant-
cultivators…and some of the sons and relatives of the landholders."[29]
He argued that Hideyoshi's surveys conveyed to peasants a security of
usufruct and assured them benefits of their investments to improve the
quality of their land and to boost its productivity.[30] Finally, John W.
Hall noted that Hideyoshi's surveys "enforced on a national scale completely new systems of land registrations, tax assessment and payment,
and land tenure"; the land surveys identified the "presumed owner" and
conveyed to listed cultivators a "security of occupancy that bordered
on ownership."[31] Thus, discussion of transformations in landownership
rights fits into the general narrative of the evolution of "economic society" that Hayami posits and with which a number of other economic
historians concur.

However, as Hayami notes, during pre-Meiji economic development, a number of prominent holdovers remained—even after the
Tokugawa era. According to Hayami, they reflected regional differences in the spread of "economic society" through the general, especially
rural, population. The large role of small-scale enterprises through
the twentieth century represents one post-Restoration, modern example.[32] The surprising persistence of small-scale farm activity in the
post–World War II era provides a distinctive case, one that Penelope
Francks argues was made possible by rural families' ability to combine
agriculture with roles in other economic activities (labor, business salary man, government employee, and so on).[33] Some of these practices
appear to be remnants from the past, emblematic of the failure of "economic society" to penetrate a sector or region. Yet, as Francks suggests,

the apparent continuity of form can mask a transformation of content. In the case of the small farmer, she notes the increased importance of women's labor in permitting the survival of small farm households down to the present day.

Like the small farm operations that Francks analyzes and the other examples cited above, Tokugawa era joint village control of access to arable land and periodic reallocation of that access survived well into the twentieth century, albeit on a considerably smaller scale. The presence of joint ownership in Tokugawa society and its post-Restoration survival reminds us that regardless of general trends toward modern forms of business and economic activity in the Tokugawa era and later, significant segments of Japanese society did not exhibit the patterns we associate with modern economic activity. Regional variation could be considerable.

The themes outlined above represent one facet of a broader treatment of Japan's rural and environmental history, histories with more diverse concerns than economic growth in the early modern era and its relationship to growth of Japan's modern economy. Spreading village literacy and a robust economic diversification spawned increasing numbers of village documents, especially from the eighteenth century, offering insights into the relations of Japanese with their environment and the functioning of their rural communities. Conrad Totman, Brett Walker, and David Howell all have addressed environmental history during this era, focusing largely on the nonfarm activities of forestry, hunting, and fishing.[34] Such activities had an impact on farming communities—forestry provided watersheds and building materials, fishing and hunting commonly supplemented food crops, and the former provided fertilizer. In addition to Smith's classic analysis of agricultural diversification, commercialization, and change in household structures, studies by William Kelly, Anne Walthall, James White, Herman Ooms, and others have explored the functioning of Tokugawa irrigation networks, village protests, and social status in Japan's farming communities.[35]

Japanese scholars have produced an extraordinary array of studies on premodern agriculture and village society, especially for the early modern period and beyond. The most well-known of such scholars in the West is probably Furushima Toshio, a diligent student of the agricultural communities of central Japan, but there are dozens, if not hundreds, of others.[36] For these scholars, landholding size and what it says about village classes in Japanese history has been of more concern than the study of landholding regimes per se.

While these studies (Japanese and English) and more have enriched our understanding of Tokugawa villages, a significant lacuna remains: how did rural communities deal with the vagaries of nature that directly determined agricultural outputs? In eras when agriculture comprised the livelihoods of the vast majority of the population and the major source of state and personal wealth, this is a significant omission. Study of joint ownership of arable land provides some answers to this question, and in the process, it encourages us to rethink the character of a farmer's relationship to the land his family cultivated.

A farmer's relationship to the land was not always that of one plot, one owner. Japan has a long tradition of communitarian control over important local resources; however, our awareness of these practices is largely limited to one ancient practice and to control of largely uncultivated common lands. Students in survey courses covering early Japanese history are introduced to an extreme form of public control of a resource when taught about the Japanese attempts to adopt the Chinese "equal field" system. This practice (the *kubunden* or *jōri* system) was central to the tax and land tenure policies known as the Taika Reforms that began in 645.[37] Under this regime, the state periodically measured lands, reinvestigated their productivity, conducted rural censuses, and reallocated land to peasants based on the size and composition of their families. Documents from this era indicate serious efforts at regular census taking, and stone markers found throughout central and western Japan today prove that authorities invested heavily in evaluating and measuring lands for allocation. However, our student also learns that the "equal field" system, like so many practices modeled on Chinese patterns, did not suit Japanese society and soon disappeared. Intriguing as it was, the "equal field" policy is typically viewed as a unique, temporary exercise of government control of arable lands.

Quite distinct from the ancient system of land reallocation, and in addition to it, a more advanced student may learn about a different practice, one that was used to manage nonarable common lands (*iriaichi*). A village, or sometimes a number of villages, exercised joint control over these lands to ensure that all people entitled to use the grasses, firewood, and other common resources had fair access. Common lands existed in most parts of Japan, and joint ownership practices on them originated in Japan's medieval era. Communities exercised control in a highly structured and sophisticated manner.[38] Farmers sometimes planted dry-field crops such as red beans (azuki) or giant white radish (daikon) that did well in poor soils. When such lands were composed of rushes or brush, common lands provided thatch for residences and green manure for fields. Such lands were sometimes used for slash-and-burn farming. Access might be assigned by lot, or participants might be allowed to harvest a certain amount of a resource or be permitted entry to harvest a resource on a regulated schedule. These rights, however, could not be accumulated or lost via market mechanisms. Village regulations determined access to resources and punishments for those who violated them.

Each of these practices, both ancient land reallocation and *iriai*, are distinct from those that concern us here. Although the former dealt with the core arable lands of villages and some scholars conjectured a link to early modern *warichi*, none has been demonstrated. *Iriai* lands were neither the economic core of village life nor the core arable lands of a community, no matter how essential they were to farm enterprises and households.[39] While we might reasonably speculate that *iriai* practices provided a template for *warichi*, we have no documentary evidence to support such a conjecture.

*Warichi* has not been widely studied by postwar Japanese scholars and, except for my previous publications on the subject, is almost entirely unknown in English-language works on Japan.[40] A relatively frequent subject of prewar Japanese historians, it fell out of favor in the postwar era. The general failure of Japanese and English-language scholarship to explore joint ownership was conditioned by two circumstances: the predominant academic interpretations of the Tokugawa state and society that emerged after World War II and issues of documentary survival.

The emphasis on nationally defined rights over agricultural land introduced above largely precluded any consideration of regular village control over arable land. The story of a national transformation of landholding practices at the behest of a powerful national hegemon provided Japanese historians something of a revolutionary event to dissect. It was indisputably important. To think of local transformations and the complexity of landholding practice, including *warichi*, suggested less than a major makeover and limitations on the political authority of the emerging early modern state. Such a story might even smack of Japan's "backwardness." None of this is to say that Japanese scholars openly denied that *warichi*-like systems existed in the seventeenth to nineteenth centuries; they simply marginalized their importance or ignored them altogether. It was into this intellectual atmosphere that American and other English-speaking scholars entered as Japanese studies expanded in the postwar era.

This intellectual environment was reinforced by the limited and opaque nature of the documentation of *warichi*. Early documentation is sparse and widely scattered. It took the preeminent postwar scholar of joint landholding, Aono Shunsui, years to develop the data for his impressive studies, even though he focused most heavily on cases of domain-controlled *warichi* that are relatively well documented.[41] Often all that remains are the notebooks that describe the final outcome of a redistribution. Made for internal use only, these can be cryptic and even indecipherable. All these circumstances discouraged exploration of *warichi* as an extended research topic. Consequently, in the post–World War II era, the modest number of articles on *warichi* are brief, opportunistic publications rather than part of a research effort devoted to thorough study of joint land tenures.

Unlike the studies of Hayami, Saitō, Francks, and others noted above, which examine individual families or enterprises, or those of Yamamura and Hall that attempt to generalize at the national level, this book explores village practices and highlights regional variation, not uniformity. We must be cautious about focusing only on the family "enterprise" in thinking about the foundations for Japan's economic transformation. As Nobel laureates Douglas C. North and Elinor Ostrom and their colleagues have shown through impressive research, the institutional contexts in which economic activities take place exert a powerful influence on the decisions people make and the course along which economies develop.[42] In Japan, the institutional framework that structured rights to exploit land was created primarily by largely self-governing villages and secondarily by baronial domains *(han)*, not by a national legal structure. In this regard, the same forces that influenced access to water for irrigation, the other major communitarian lever influencing agricultural development, also shaped the structure of rights to land. This means that if we are to understand the development of Japanese landholding practice as a whole, we must create interpretations that encompass variety rather than impose uniformity. This caveat holds across time as well as space. Despite the sharply different political character of the Tokugawa and post-Restoration regimes, many scholars paint a similar picture of central governments capable of creating nationally uniform landownership rights, through Hideyoshi's surveys in the first instance and via the new Meiji Land Tax Reforms of the 1870s in the latter case. (Customary and communal property rights [*iriai*] were recognized in principle, and some lands, a number of domain forests, for example, were treated as public lands under the Meiji reforms.)

Ironically, and directly apposite to the question of diverse versus uniform practice in landholding, the overwhelming majority of scholars who have studied *warichi* tie its creation to the emergence of new political arrangements under the early modern state of Hideyoshi and his Tokugawa successors. At the same time at which some see the genesis of more secure private ownership rights, rights in land that scholars characterize as "nearly modern," joint control of arable lands arose in widely scattered parts of Japan (especially in the Hokuriku, Shikoku, and Kyushu). As I discuss below, these scholars see the new late-sixteenth- and seventeenth-century structures of land taxation as a critical factor in stimulating the growth of *warichi*. They link development of village responsibility for land tax payment and errors in domain land surveys that established the basis for tax assessments as the stimulus to reallocation under joint land ownership.

That scholars could come to such different evaluations of the role of Japan's late-sixteenth- and early-seventeenth-century land surveys, one stressing their ability to create fee simple ownership, the other stressing their role generating systems of joint village ownership of arable lands, is just one of the conundrums that the preceding discussion raises. It raises parallel questions for Japan's late-nineteenth-century transformation, in which new systems of land taxation supposedly made all ownership of arable fee simple: How could joint ownership survive the formidable Meiji legal onslaught? In light of interpretations of the role of fee simple ownership in promoting economic growth and preserving natural resources, why would any community want to continue to practice joint ownership in early modern Japan? What were *warichi*'s functions, and why did it survive changing economic environments? If continuation of this practice during the Tokugawa period is puzzling, how much more fascinating it is that we find cases persisting into the 1970s! Is joint ownership adaptable, or are the challenges of a modern economy less forceful in eliminating these arrangements than we have thought? Do areas that practice *warichi* experience a degradation of arable land resources? Scholars often stress nationally uniform tenure systems, both early modern and modern, so what enabled regional variation?

Joint control and periodic reallocation of cultivation rights contradict standard images of intense attachment of Japanese farmers to specific pieces of land (in Japanese, *tochi aichakushin*).[43] Scattered throughout Japanese local histories and documentary repositories, extensive early modern and modern evidence records intense disputes over possession of individual families' land and strongly supports the validity of this image. Even in areas that practiced *warichi*, records fill the local archives detailing disputes over land, prices paid, forfeiture of cultivation rights for failure to repay loans, and terms for return of rights to cultivate land that were in principle transferred forever to others.[44] How are we to interpret such documentary evidence in the *warichi*-practicing regions where attachment to specific plots of land was impossible?

The chapters that follow explore these and related questions that bear on our understanding of both Japanese social, economic, and political history as well as conceptualizations of social science theory regarding the nature and role of property rights in society. My conclusions ask future researchers to be cognizant of subtle and varied influences on the development of *warichi* and other practices in Japanese society. My effort here challenges scholars to embrace the varied roles of actors at multiple levels of society, not just political leaders at the top or at the bottom of the political hierarchy. At a broader theoretical level, it makes a case for more subtle differentiation of property rights in land. In both the theoretical and historical realms, my work tests common generalizations and suggests alternative understandings.

Let us now turn to the fundamental question of the origins of *warichi*. The origins of these landownership regimes remains cloudy. Although Japanese scholars have offered a number of explanations, each has limitations. A review of Japanese scholarship will provide a foundation for some of the hypotheses about *warichi*'s origin, characteristics, and effects that I test more fully and examine critically in the chapters that follow.

# 2

# Origins and Geopolitical Contexts

Chapter 1 outlined several intellectual contexts in which the study of joint village control of arable land is important; however, significant additional questions about joint landholding derive directly from the explanations that Japanese scholars have offered for its origins.[1] This handful of explanations raise several common issues. First, with one exception, these explanations pose questions about the relationship between systems of joint landownership and the natural environment in which they existed. Second, by implication they raise issues associated with the geographic distribution of *warichi*. Third, they raise questions about what permits or encourages this distribution, in particular the historical administrative contexts in which joint ownership develops. Consequently, this chapter reviews key geographic conditions and considers the Tokugawa administrative context both nationally and for the region of our key case studies, Echigo.

## Theories of Origin and Their Implications

For the most part, the Tokugawa era is seen as the origin of the various joint forms of landholding to be explored below. While there is general agreement on when they first appeared in Japan, at least superficially there is little agreement on why the *warichi* system originated when it did or on its predominant function. With the exception of a single dissenting view, early scholarship on *warichi* established that while evidence exists for joint control over common lands *(iriai-chi)* prior to the late sixteenth century, no such evidence exists for its use on arable lands. Up to the time of Oda Nobunaga and Toyotomi Hideyoshi, rights to the benefits of arable land were multilayered, a remnant of the ancient *shōen* system that shared rights to the income from the land not only among tenant, cultivator, and resident landlord, but also with parties far removed from the land—powerful individuals in provincial capitals, large temples and shrines, and court or military figures. Joint ownership of arable land appears roughly coincident with the end of these land rights and the emergence of villages as largely self-governing units.

Early studies also established that joint ownership of agricultural lands was highly varied in its content, making generalization a challenge. In a given village, the practice might be applied only to seedbeds, or just to mountain fields, or to all of the arable land in a village. There was also significant variation in the period that elapsed between reallocation of cultivation rights. The mechanisms for effecting redistribution and customary names for the practice varied widely.

In practice, there was a range of community rights in land evident in Tokugawa villages. At one end of the continuum were those villages that exercised joint community control over village common lands but not over ordinary farmlands. In the central portions of the continuum were villages that redistributed parts of village arable land in addition to the joint community rights exercised over common land. Finally, at the other end of the continuum were those villages that applied redistribution practices to all of the land in a village.

All of these forms of landholding coexisted in Tokugawa Japan. In the late sixteenth and early seventeenth centuries, management of the commons reached its maximum sophistication. This development coincided with the appearance of joint ownership of arable and rights akin to fee simple ownership.

The most persistent theme in explaining the origins of joint land tenures lies in the geographical circumstances of villages, especially in the case of the most common form, what I refer to as per share joint ownership. That said, modern scholars have cast their arguments so as to suggest distinct theories of causation and do not state explanations in baldly geographically deterministic form. Scholars widely emphasize the importance of corporate village responsibility for payment of domain taxes as a significant and necessary condition for the rise of land redistribution practices, but they do not see this alone as a sufficient stimulus to adopt joint landholding on arable lands. In the context of corporate village payment of land taxes, they view *warichi* as one device for distributing the land tax within a village.

At the height of interest in *warichi*-like systems in the 1920s and 1930s, the eminent rural historian Furushima Toshio grouped theorists of the origin of land redistribution into several categories.[2] First were scholars who argued that the *warichi* system was a remnant of the seventh-century Taika era "equal field" (*jōri* or *kubunden*) system of redistributing lands based on the Tang Chinese model: allotting lands to households according to family size and composition. Most historians, however, quickly concluded that this argument was unverifiable given the lack of evidence connecting this system to seventeenth-century and later practices. All evidence points to its complete extinction by the late Heian era.

A second theory explained redistribution as a result of frequent floods and landslides that destroyed arable lands. This interpretation has a long history and was, for example, taken up in the well-known manual of Tokugawa administrative practices *Jikata hanrei roku*.[3] Proponents agree that villagers in flood- and landslide-prone areas used joint ownership to spread the loss from such disasters among all holders of land rights. In effect, this framework treats *warichi* as a form of insurance or risk diversification. If this were in fact the case, one would expect to find joint ownership quite routinely, at the village level if not at the level of domain policy, given Japan's widespread vulnerability to flooding and other risks. One would also expect to find it on both long-standing, base fields (*honden*) as well as reclaimed land (*shinden*) and that it would be present regardless of tax regime.

Third, Nakada Kaoru and others argued that *warichi* systems arose from cooperative development of new agricultural lands.[4] Existing rice paddy agriculture and its associated technology could not create fully productive fields in the first year of cultivation. Creating an impervious pan that would not lose water during the period of irrigation, proper conditioning of the land, and so forth all took years. Early modern villagers and domain authorities alike typically presumed that it would take five years for land to reach its full, stable productivity. Consequently, newly cultivated land was reallocated annually during the reclamation project to all entitled. At each stage of reclamation projects, the rotation of lands assured equal rewards for those who had made comparable investments in the project. No one family had access to all of the best land. The same logic applied when dry field was converted to paddy. If this reasoning is accurate, joint ownership should have been implemented almost exclusively at the village level, with little domain involvement, since reclamation and land improvement over the course of the era was largely a village affair. Furthermore, it would appear only on new and improved fields. Since the productivity of lands ultimately stabilized, joint ownership should appear as a temporary measure and then cease when lands matured, and it would generally affect only a limited part of a village.

Fourth, Makino Shinnosuke stressed the relationship between the villagers and the land tax system. He linked the origin of *warichi* to the establishment of village responsibility for tax payment. In villages that had multiple overlords, for example, when several of a daimyo's retainers each independently assessed and drew revenues from a village, different families within the same village were potentially subject to different tax rates from independent assessments of yield.[5] The *warichi* system averaged the tax rate for the entire village. According to Makino, peasants also developed the *warichi* system to equalize the land tax burden among their numbers in the face of changing fertility of the fields. If true, one would expect the system to be very widespread, especially in the early seventeenth century, when multiple overlordship over villages was common. It would appear first and foremost at the village level and would disappear with the end of the Tokugawa system of corporate village payment of land taxes.

The last three of these theories touch on the recent debate as to the character of peasant economic decision making, the debate between the "moral economists" and the adherents of the "rational peasant" thesis. At the risk of oversimplification, one might say that the "moral economists" stress peasants' low-risk, security-conscious attitudes that encourage maintenance of the their current standard of living and the patterns of mutual obligations that peasants bear as a result—characteristics that superficially appear to underlie *warichi*. Adherents of the "rational peasant" thesis emphasize peasants' drive to increase their standard of living through individual, household, and village investments potentially antithetical to *warichi* practice.[6] The emphasis on fairness in taxation in *warichi* reflects a position closer to that of a moral economy than a model of individual villagers exercising rational economic choices.

Aono Shunsui, author of the most recent large-scale study of *warichi*-type systems, also bows toward a moral economy perspective and views the appearance of such systems as a consequence of conditions unique to the Tokugawa era—the development of village collective responsibility for land tax payments, the use of general village economic conditions (not just the income from arable lands) in setting village tax assessment value (*kokudaka*), reclamation conditions, and other factors.[7] Through a series of well-executed studies of joint landholding in different parts of Japan, Aono argues that *warichi*-type systems were the final product of a range of mutual aid measures, implemented as circumstances warranted, to deal with the hardships of both farming and land taxes.[8] Central to his discussion is the appearance of a gap between documents created by domain authorities to calculate a village's land tax base, and what the villagers calculated to be their actual ability to produce crops to pay taxes. His argument presupposes that either samurai officials were incapable of accurately assessing the land's productivity or they conducted land surveys so hastily that their records contained widespread inaccuracies. This gap spurred villagers to conduct their own surveys and to allocate land rights in such a way as to create a reasonably equitable and fair allocation of taxes. Although these early-seventeenth-century cadastral surveys stimulated creation of joint landholding, the long-term reasons for the gap lay overwhelmingly in the realm of changes resulting from (a) floods and landslides that destroyed the agricultural utility of land but did not necessarily result in their delisting from the tax registers and (b) addition of newly reclaimed land that would affect villagers' sense of fair distribution of land tax but for which adjustments in a village's official putative yield might not be made in a timely manner, owing, for example, to deliberate concealment of the reclamation from authorities.[9] In sum, while this interpretation limits the role of natural disasters, flooding, landslides, and so forth in stimulating the development of redistribution practices, it does not eliminate it, and continuation of joint ownership well after realmwide domain land surveys was linked to frequent changes in farming conditions. If this interpretation is valid, the origin of the practice of *warichi* lies in villages, not in domain policy. In this case, the practice would not be universal but would be common and would disappear with the implementation of the Meiji tax regime that focused on individual land tax payment. It might also be expected to decline if better, more accurate records were created over time.

Despite the delineation of apparently distinct stimuli to the creation of joint ownership, upon reflection, the preceding explanations isolate only one distinguishing characteristic that poses a hypothesis for further investigation: *warichi* is found where natural risks (e.g., floods, landslides) are high. Most of the implications related to the geographic distribution of joint ownership are not evident in the historical record, and many of the purported causes do not distinguish joint ownership areas from non–joint ownership regions: Corporate village responsibility for assessing and paying domain taxes was typical of all Japan from the late sixteenth to the mid-nineteenth century. Listing of individuals did not establish any family's obligation to pay taxes directly to the domain; instead, land taxes in all Japan were routinely assessed on villages, not individuals. Therefore, this factor does not help to distinguish those areas of Japan that employed joint ownership from those that did not. Since joint ownership continued long after a reclamation project had been completed, the need to allocate access to newly reclaimed land in proportion to participant inputs also appears to be an inadequate explanatory factor. The fact that new lands brought under the plow during the early modern era were overwhelmingly in marginal natural settings—locations subject to flooding, poor drainage, landslides, irregular water supply, and the like—suggests that unstable geographic conditions were an important factor in the long-term use of redistribution on reclaimed land.

The preceding reasoning brings the weight of scholars' previous explanations back to natural conditions as the chief stimulus to create joint ownership practices. A key task below is to assess the degree to which the presence or absence of *warichi* and its operating practices is in fact tied to natural hazard risk (see Chapter 6 in particular). The situation turns out to be complex, both because of diversity in *warichi* practice and villager interpretations of the nature or severity of threats from natural hazards.

The preceding discussion has made three key points. First, while scholars have looked at redistribution practices in a variety of contexts and while they have offered differing explanations for the origins of joint village control of arable lands, the common distinguishing thread explaining its presence and continued use lies in variations in the natural conditions of agriculture and variations in the amount of land under the plow. (The principal exception is the explanation that relies on adoption of Chinese practice.) By implication, a task for the following chapters will be to assess the degree to which variations in natural environment account for *warichi*.

Second, the discussion on village responsiveness to local natural conditions suggests that the political organization of Tokugawa Japan fragmented control over land rights. Although I discuss below cases in which regional domain control was prominent, villagers themselves often determined what rights in land were vested in whom. No political structure existed through which national policies regarding landholding could be enforced.

Finally, this overview carries implications in one other area of general interest—the role of the Tokugawa village in creating the institutions of local government. Especially in the context of *warichi* origins, it reveals creative village contributions to local administrative practices. Rather than being the passive recipient of daimyo-conceived administrative arrangements, villages created programs that, as I examine below, domain authorities later embraced.

Diverse loci of decision making set the stage for a key characteristic of the following analysis: the lack of standardization of rights in land makes parsimonious generalization about any land system difficult. The wide variation in land tenure regimes is evident even within the narrower context of joint forms of landholding and management. However, one can identify significant parameters of interaction, which I present in Chapter 4.

The national geographic and political contexts in which early modern joint ownership developed bear directly on the three themes just highlighted and provide a foundation for a classification of joint ownership regimes. I explore them next. In addition, before moving on to an analytical classification of joint ownership systems and an assessment of their extent throughout Japan, I sketch the geographic, economic, and political contexts of the Echigo region, site of my most extended case studies.

## The Japanese Geographic Context

Japan is a heavily mountainous country, built from volcanic and tectonic activity. *The National Atlas of Japan* classifies 75 percent of the country as mountainous. Peaks rise at steep angles, although they are not extremely high. Fuji, the highest peak as well as Japan's most famous, is some 3,800 meters (12,400 feet) high. All mountains extending beyond the tree line (2,500 meters) are in central Honshu, roughly the latitude of Tokyo. While news media remind us of Japan's volcanic base via spectacular images of active volcanoes and footage of Japan's tragic earthquakes, we are much less aware that landslides and floods are frequent and substantial. In Japan, evidence of both is widely apparent: traveling along Japan's highways and railways, many, many slopes are covered with wire netting or cement, or concrete blocks are bolted into the hillsides, all in efforts to minimize landslide damage.[10] Heavy or persistent rain and earthquakes are major triggers for landslides, but so, too, are road, rail, and related construction. At base, regardless of the trigger, much of Japan is subject to landslides because of the geological youth of the land. Many of its surface soils are quite young and unstable, especially in central Japan, the area in which multiple tectonic divisions come together to form the *fossa magna*.[11]

Mountainous terrain precludes the great shifts in river beds such as the north-south movement of the Huang He (Yellow River) in China—movements that historically cost many thousands of lives. Nonetheless, many parts of Japan experience frequent flooding on a smaller scale. This heritage shows in place names like Oshimizu—literally "pushy water"—a town in Ishikawa Prefecture, and many others. Local histories tell of repeated and serious flooding.[12] Downstream communities in narrow valleys are exposed to risks similar to those posed by flash floods: residents of Tokamachi in central Niigata Prefecture understood that even if they themselves were not experiencing rain, they might be flooded if heavy rains fell upstream in Nagano, many kilometers away. The mountainous nature of Japan's topography constrains the length and depth of its rivers. It has very few long rivers, and most are of inadequate depth for deep-bottomed transport, even by Tokugawa shipping standards. Consequently, there is widespread potential for moderate flooding, which affects not only agricultural lands, but the residential hamlets lining steep slopes and small floodplains.

The letters of late-nineteenth-century English traveler Isabella Bird bear out the ferocity of Japan's rivers and streams. While in northern Japan in August 1878, she endured days of rain on end:

> As I learned something of the force of fire in Hawaii, I am learning not a little of the force of water in Japan.... The hillside and the road were both gone, and there were heavy landslips along the whole valley.... The rush of waters was heard everywhere, trees of great size slid down, breaking others in their fall; rocks were rent and carried away trees in their descent, the waters rose before our eyes; with a boom and roar as of an earthquake a hillside burst, and half the hill...was projected outwards, and the trees, with the land on which they grew, went down heads foremost, diverting a river from its course.... The Hirakawa, which an hour before was merely

a clear, rapid mountain stream, about four feet deep, was then ten feet deep.... On thirty miles of road, out of nineteen bridges only two remain, and the road itself is almost wholly carried away![13]

Some of what Bird witnessed may have resulted from seventeenth-century deforestation, but in fact such flooding, like landslides, has a long history. While well-vegetated slopes may alleviate flooding and landslides, the protection has limits. Many landslides have their root in the interstices between soil layers that lie well below the depth of plant root systems, and flooding depends on a variety of factors, including the level of the water table, previous saturation, and soil type. Massive soil deposition formed a number of Japan's plains within the past six millennia, long before human reengineering. Indeed, premodern human engineering—dikes, drainage, and water storage—was limited in scope.[14]

Maps of Japan seen by most of us in the West oversimplify Japan's topography, hiding the detail that suggests the risk of floods and land-slides. Figure 2.1 shows a fairly typical topographical representation of Japan, clearly indicating the extent of mountain coverage and the restrained size of its plains. Only the Kanto Plain, in which Tokyo is situated, appears as a sizable, flat area. For Americans and many Westerners, maps like this are likely to convey an image of the Kanto and other plains as being flat as the Kansas prairie.

However, such an impression is misleading. Maps, like essays and books, generalize. They simplify as an aid to understanding, but often we are oblivious to this process. We compare maps explicitly or implicitly without considering the different mechanisms map authors employ to render a comprehensible image.

Map images manipulate our understandings in multiple ways. Manipulation of scale can radically change our impressions of topography. A comparison of Figure 2.1 with Figure 2.2 and many of the other maps in this book shows the degree to which small-scale national maps like Figure 2.1 conceal more than detailed, larger-scale maps. Thus Figure 2.2, which uses the same 1-kilometer-mesh elevation sampling as Figure 2.1, shows more detail owing to a change in scale, that is, one can see greater variation in elevation around the core of the Kanto Plain, yet the plain still seems quite flat even when rivers are added.

Methods for representing elevation represent another significant manipulation. Even without a legend of elevation, Figure 2.1, present-

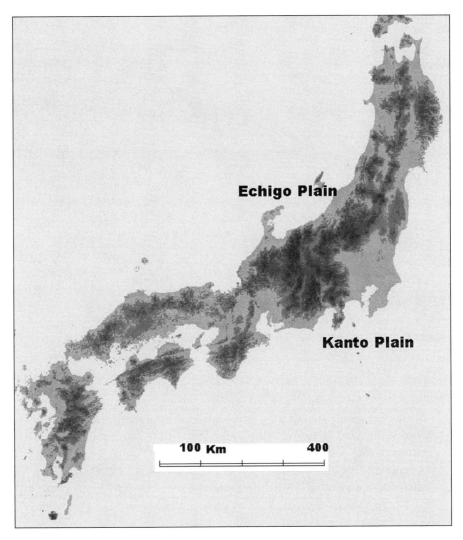

**Figure 2.1**     One-Kilometer Mesh Relief Map of Honshu, Shikoku, and Kyushu

ed here in grayscale, clusters elevations into a few ranges to highlight Japan's central mountain backbone and its less undulate regions. This process ultimately distorts by rendering the latter as flat.

While most people may recognize distortions of scale and representation of elevation (among other distortions common to traditional cartography), the dawn of digital cartography highlights another source of distortion: the method of calculating elevation. Given my use

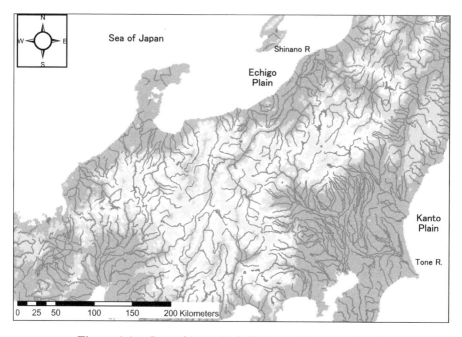

**Figure 2.2**    Central Japan Relief Map, 1-Kilometer Mesh

of digital mapping, this detail deserves some explanation. To create the base for a digital map, elevation models are made by sampling elevation points systematically in a grid pattern. The greater the distance between sampled elevations—the mesh size—the coarser the digital elevation model (DEM). The elevation samples in Figure 2.1 are about 1 kilometer from each other, or a 1-kilometer mesh. One outcome is that someone familiar with more precise maps of Japan will recognize distortions in the coastline as presented in this map. Peninsulas seem particularly distorted. In addition, this coarse sampling provides a less refined model of variations in elevation than a sampling using a 250-meter or 50-meter mesh.

Representation of variation in elevation is critical for understanding environmental influences on Japanese farming villages: despite distortions of elevation and the coastline, for example, Figure 2.2 is helpful in drawing attention to two important issues pertinent to a consideration of joint landownership. First, it clearly illustrates the character of Japan's rivers with many short streams cascading down steep slopes. Along two of Japan's major rivers, the Shinano and the Tone, what stands out is the large number of small valleys and streams running between low mountains and hills in the Kanto area and the Echigo Plain, where the Shinano River empties into the Sea of Japan.

Second, Figure 2.2 helps frame a question regarding the distribution of joint ownership of arable land. Neither the Kanto area as a whole nor the vicinity of the Tone River in particular is thought of as a region where joint ownership was practiced; the Shinano drainage basin, however, is considered a classic region for *warichi*. Both regions are characterized by a network of rivers that drain large areas of Japan, ultimately via two of Japan's longest rivers. The difference between the two areas in climate, precipitation, or other natural conditions is not immediately evident; nothing leaps out to explain why one area would have used joint ownership extensively but the other did not. There are differences, but they do not appear to be dramatic. Annual precipitation is higher for the western part of central Japan (over 78 inches) than its eastern counterpart (59–78 inches), and there is much more precipitation in the winter to the west than in the east. Approximately similar circumstances, very different landownership regimes. Myriad factors combine to create natural hazard risk, and no firm conclusions can be reached at this broad scale of analysis, but even at this level, the issue can be posed: do similar geographic circumstances correlate well with the presence or absence of *warichi?*

## Early Modern Japan's Administrative Diversity

While popular images of the early modern Japanese hegemons and even many scholarly treatments portray an imposing, at times terrible, and powerful authority capable of rigorously suppressing rebellion and moving domain lords from one part of Japan to another, more recent English-language scholarship stresses the limits of shogunal authority and the independence of daimyo domain administrations in the years between 1600 and 1868.[15] The varying positions that scholars take depend in part on the evidence they analyze. Those who emphasize the shogun's power largely focus on analyses of documents emanating from his administration; those who focus on the limitations of shogunal authority have relied on domain documents. While there is room for disagreement and debate on this issue, the discussion of the Tokugawa administrative setting below is based on well-recognized limitations on Tokugawa shogunal and even domain authority.

First, there was no extensive, integrated, Japan-wide system of domestic civil administration atop which the shogun perched. While

Oda Nobunaga (1534–1582) and particularly Toyotomi Hideyoshi (1536–1599) are widely seen to have increasingly strengthened an emerging central political authority that was inherited by Tokugawa Ieyasu (1543–1616), these men and successor shoguns typically could not treat daimyo domain administrations as bureaucratic subdivisions of shogunal administration. Shoguns could demand irregular financial contributions to projects such as the rebuilding of Osaka Castle after the final defeat of the Tokugawa opponents (1615) and other personal service, typically military in nature.[16] They could adjudicate disputes that crossed domain boundaries. However, they were incapable of implementing any regular system of taxation on either domains or villagers throughout Japan. Taxation, valuation of land for tax purposes, and the like were the provenance of daimyo—a role that Oda, Hideyoshi, and the Tokugawa shoguns played solely vis-à-vis their own privately controlled domains.

Shogunal ordinances could set patterns that daimyo might choose to imitate, but no centrally directed policy related to landownership could be compelled.[17] Despite the bombast in Hideyoshi's orders and the high-sounding principles of shogunal edicts, no effort to enforce such policies within domains was attempted. This meant that daimyo could exercise their own judgment as to how to structure policies toward landownership, and exercise it they did.[18]

In the absence of a hegemon's ability to command bureaucratic obedience within daimyo domains, daimyo administered local affairs with a high degree of autonomy. The number of daimyo domains varied over the course of the era; there were approximately 260 to 280 at any given time, a number that suggests the potential for multiple permutations in local administrative patterns. Daimyo domains varied widely in size, from the largest like Kaga and Satsuma, which each occupied several of Japan's sixty-six traditional provinces, to subprovince domains as small as half a county (*kōri*, modern *gun;* typically comprising five hundred to six hundred villages).

Divergent from shogunal patterns or not, the degree to which daimyo might effectively pursue a policy was influenced by the degree to which each domain consisted of contiguous territory. While virtually all daimyo ruled territories that were scattered in some measure, bigger domains tended to be composed largely of contiguous territory, and smaller domains controlled a much higher percentage of scattered territories.[19] Even the largest domains had their main holdings interlaced with small scattered holdings of the shogun, other daimyo, and shogunal rear vassals *(hatamoto)*. Likewise, a large domain lord would have some territory located many miles distant from his main holdings. Nonetheless, such domains can be treated as unified by comparison with the majority of domains. Distance and the transportation methods of the day combined to compromise the ability of a smaller daimyo to effectively impose his will on villages in a number of scattered regions of Japan. Indeed, even in a large contiguous domain, the challenge of distance could be considerable, as in the case of village-by-village inspections and crop samplings *(bugari)* to determine per hectare yields to use in land tax assessment. The task was time-consuming and time-constrained; villages all had to be inspected at a time reasonably close to but before the harvest, a constraint that had the small number of officials scampering around the countryside, rushing to complete their task expeditiously.[20]

Even within a relatively well-controlled domain, key elements of local (village and town) administrative decision making, including policies regarding landholding, resided in commoners' hands. In rural districts, a village headman in concert with a village council ran each village, not a samurai.[21] (As a rule, samurai were prohibited from residing in villages or owning land. While there were exceptions based on daimyo policies that varied over time even within a domain, this generalization applies to most of Japan and most of the Tokugawa period.)[22] Even subcounty district administration lay in the hands of commoners.[23] Generally speaking, while there were samurai officials *(bugyō)* in charge of county-level administration, these men, like most samurai administrators, were concerned primarily with maintenance of public peace, collection of taxes, and coordination of those few areas of administration that required the cooperation of multiple villages or districts, such as riparian works and allocation of irrigation water.[24]

Domains were often indifferent to land rights structures if taxes were paid, a circumstance that permitted diverse practices from village to village, even within a domain. This is true for the vast landholdings of the shogun, where some villages practiced *warichi* and many others did not, and even holds for small domains, like Tōdō, discussed below, where we see a similar division in village landholding practice.[25] Consequently, hasty generalization about landholding forms, practices, and policies is all but impossible.

While many early modern Japanese history specialists will recognize the limitations on state authority I have noted above, they may be unfamiliar to others. The reason lies in the failure of widely read general treatments to convey concrete examples of the consequences of domain administrative autonomy, this despite recent monographic studies and some periodical literature pointing to the high degree of administrative latitude with which daimyo acted.[26] The most direct discussion of the problem of characterizing the Tokugawa state appears in Marius B. Jansen's text: in his view, the Tokugawa political order exhibited such a high degree of decentralization that he questions whether or not the Tokugawa political order even constituted a "state."[27] Of course, to the degree that the Tokugawa conducted foreign policy, adjudicated disputes between daimyo, and so forth, it did function as a state, but Jansen's evaluation is a stark reminder of its limitations as a national administration.

To some degree the challenge of describing to general audiences a complex political order in a short space—whether in a survey text or as background to treating some other subject—restricts what authors say, but absent clear examples of independent domain actions, shogunal strength remains the dominant image. Unless the author's subject specifically hinges on variation in administrative structure and practice, this element of Tokugawa politics is elided. The maintenance of long-term peace reinforces impressions of strong central authority since it is hard to imagine maintaining public order and political autonomy without such a government. Thus, while monographic treatments increasingly demonstrate the role of local initiative, one can still get the impression that shogunal leadership and initiative were paramount in local administrative practice.[28]

While great domain autonomy was the rule, village-domain relationships throughout Japan commonly shared a significant feature: daimyo assessed taxes on each village as a corporate entity rather than taxing individual farm households. Primary responsibility for distributing domain tax burdens to individual households within a village lay with resident village leadership. Villagers employed a wide array of devices of varying complexity for distributing domain taxes among themselves. This said, on occasion, domain policy addressed allocation of taxes within villages in an effort to assure that there was a modicum of equity and fairness, attempting to eliminate the worst inequities that might lead to a loss of agricultural labor due to bankruptcy or disputes that threatened public disorder and challenges to elite rule. Policy makers' concern was with efficient tax collection and, in the extreme, heading off intravillage disturbances *(ikki)*. Noteworthy among domain efforts to encourage fairness in tax allocation and assure full payment were those that sought to regulate landholding fragmentation in order to retain farmers on village tax rolls.

The key, widely known example of administrative interference with landholding rights was widespread prohibition on the sale of land. This policy is typically explained as an outcome of shogunal decision and the prohibition of land sales in 1643 (the *ta-hatake eidai baibai kinshi rei*). However, domains acted at their own pace, and some, like Kaga, enacted similar laws much earlier (1615) and others, later.[29] Regardless of the order in which they implemented these laws, daimyo typically attempted to enforce such restrictions.

Despite the widespread presence of such ordinances, scholars generally recognize that they were extraordinarily ineffective. Villagers quickly found ruses under which they could effect land transfers. They employed adoption of an heir (from the family "buying" the land) or pledged land as security on a loan and forfeited it for nonpayment. Village documents typically record the outcomes of such transactions, so there was no attempt at cover-up.

The inability to prohibit the sale of land is a reminder of how difficult policy enforcement was at this time, even when there was widespread agreement throughout daimyo leadership. As already noted, shogunal policy regarding land could not be enforced nationally, and there was even variation within the shogun's directly administered domains. Diversity in domain policies resulted in nationally varied land rights structures and policies, not uniform practice. In addition, the difficulty in enforcement and the different degrees to which domains sought to impose a standard policy left ample opportunity for villagers to establish their own practices.

## Echigo Geography

To this point, my emphasis has been on broad overviews that bear on joint landownership, emphasizing the national geographic and political contexts in which it developed. However, to answer many of the questions raised above we must explore local cases. The most intensive studies below focus on the province of Echigo. Figure 2.3 shows the region of concern. Early modern Echigo comprised the modern prefecture of Niigata, exclusive of Sado Island.[30]

Political control in the province from the late sixteenth century was both fragmented and fluid. Appendixes of the historical dictionary *Shinpen Nihonshi jiten* list twenty different domains for Echigo Province

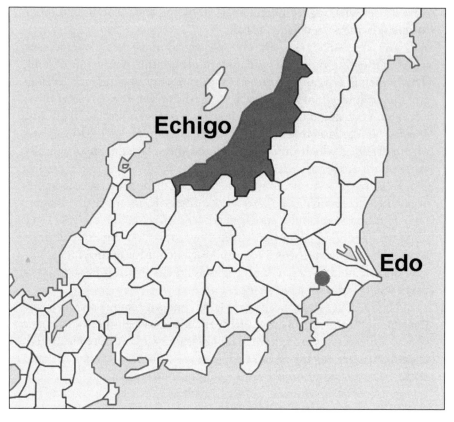

**Figure 2.3**    The Location of Echigo Province

over the course of the era. Not all were in existence at the same time, and this list does not include *tenryō* (directly controlled shogunal lands) or lands granted to daimyo outside the province either as part of their lands or to control temporarily. Some, like Itoigawa domain, came into existence, were then abolished for a number of decades, and later were resuscitated. Even when domains remained constant for a time, their boundaries might change—a phenomenon particularly notable in the area of modern Niigata City.[31]

As if these political movements were not sufficient challenge to administrative stability, any overlord—daimyo, shogun, or *hatamoto*—conducted administrative reforms of his own from time to time. Such reforms involved changes in supervisory authorities, revising the duties of village headmen or village group heads (e.g., *ōjōya*), or reconstituting the membership of subcounty districts comprising multiple villages (sometimes called *kumi*). In a number of instances, the reforms were designed to limit the potential for corruption by shifting jurisdictions and preventing the establishment of long-term relationships between governing officials and their subordinates.

However effective for other purposes, these reforms, in combination with the relative instability of domain boundaries, made it very difficult for any supravillage authority to exert consistent control over internal village affairs. This was a perfect environment for villages to operate with a high degree of autonomy. Even had domains wanted to exercise control over landownership rights, this highly fluid administrative environment created a major challenge to effective implementation.

On the whole, however, domains took no such initiative to control the nature of landownership rights. Villages operated with great latitude when it came to such rights. It was in this context that village-based *warichi* flourished in Echigo.

The political fragmentation of the province had clear economic and demographic implications. Unlike its neighbor to the southwest, Echigo lacked a single, dominant castle town such as Kaga domain's Kanazawa. Instead, it had many smaller castle towns scattered across the land. This meant that it lacked the concentrated, wealth-dispensing urban economic magnets of large, contiguous domains. For the southern and coastal regions of the province, Kyoto and the Kansai were significant economic contacts; for northern and many inland areas, the connection to Edo was stronger, and these external markets rivaled local demand.

For farmers, the lack of a large urban economic magnet meant that opportunities for diversification into cash crops were limited relative to more heavily urbanized parts of Japan. So, too, were opportunities to participate in diverse by-employments. Over the course of the Tokugawa era, such opportunities increased but within a limited scope. Crepe and other textile manufactures expanded in Tokamachi, Muikamachi, Nakazato, and other towns. Charcoal kilns were widespread. But there was no single, dominant special cash crop to compare with indigo (e.g., in Awaji), cotton (as in the Kinai), or sugar (found in Satsuma and the Ryukyus). Coastal shipping, especially with the development of the Western Route *(nishimawari sen)*, offered some opportunities for trade north and south with towns from Hokkaido to Kyushu, along the Sea of Japan coast to the Inland Sea. From the eighteenth century on, the production of rice wine *(sake)* provided commercial opportunities for one of the region's staple crops, rice.[32]

Yet even in this relatively low-demand context, Echigo farmers sought to increase crop yields. Development of local varieties of rice proceeded apace with other regions of eastern Japan. So, too, did expanded use of diverse natural and commercial fertilizers as well as new varieties of hoes, plows, and other tools. Echigo farmers not only read the most widely known of the agricultural treatises *(nōsho,* e.g., *Nōgyō zensho)* from the eighteenth century, but they composed their own, based on local experience: *Kafū nōsho, Fuzokuchō, Nōka nenjū gyōji, Ryūryū shinku roku,* and *Nōka kokoroe.* In this region, as in other areas of the country including Satsuma and Kaga, increases in yields were accomplished largely through added investment of labor. New seed varieties, improved irrigation, and flood control also made major contributions.[33]

Beyond such qualitative indications, it is difficult to get a good quantitative grasp of Echigo yield trends. Early Meiji statistics are of debatable quality; however, the change in land tax rates that took place between the late Tokugawa and the implementation of modern land taxation under the new Meiji government reflect a combination of improvements in yields and increases in area cultivated, both of which suggest the degree to which incentives to invest in land operated. Under the Tokugawa system of land taxation, taxes were levied based on estimates of crop yields for the cultivated area of villages, but scholars widely agree that the estimates on which taxes were based did not keep up with increases in actual yields during the eighteenth and nineteenth centuries. In addition, the Meiji reforms remeasured area cultivated, eliminating the common downward adjustments (e.g., for shaded lands) to area common in measurement of lands.[34] While the new system of taxation was based on an estimate of the market value of land, not yields, market value reflected the productivity of the land. A comparison of the two levels of taxation suggests the degree to which both yields and area under the plow had increased over the Tokugawa era, often without being captured by domain authorities.

At the national level, the overall tax rate was stable as the new British-inspired land tax system went into effect. There was considerable regional variation, however. The Echigo/Niigata region, along with most of Japan's northeastern region, supported increased tax obligations, while other areas saw their taxes reduced. The rates for the region suggest that pre-Restoration yields (and therefore the value of arable land and its yields) and area under the plow had increased significantly compared to the base estimates of land productivity that underlay late Tokugawa land taxation.[35]

Trends in farm family size and average farm enterprise scale for the Echigo region parallel trends scholars have identified for Japan. That is, family structure tended to become more clearly nuclear and certainly smaller over time, generally sloughing off most of the dependent labor that would have typified families in the seventeenth century and relying more on short-term labor contracts or day labor. This trend was attenuated in the more mountainous parts of the province. Finally, the average size of landholdings a farmer managed also declined over time.[36]

This said, Echigo also hosted some of the largest landlords in Japan, often interpreted as a sign of the province's social, if not economic, backwardness. These were not the famous One Thousand Hectare Landlords of the late nineteenth and early twentieth centuries.[37] The largest Edo period landlords, like the Satō family, were typically about half as large as their twentieth-century brethren, and many of these wealthy families did not develop into Meiji superlandlords. Indeed, most did not fare particularly well under the new Meiji government and its modern system of taxation.

Geographically, inland Echigo is marked by mountains, with the highest elevation reaching above 2,500 meters (Plate 1). The Shinano, Japan's longest river, dwarfs Echigo's others and collects waters from a large, complex network of streams as it flows from Nagano (the lower left corner of the map) to Niigata. In contrast, most of the remainder of the province is dominated by much smaller rivers. The *River Handbook* indicates that Niigata has more than 5,360 kilometers of rivers, more than all but Hokkaido (15,370 km) and Nagano (7,059 km).[38]

To say that the rivers and streams are smaller is not to imply that they were tame. For example, the Agano River, which empties into the Sea of Japan at Niigata just north of the Shinano, has historically presented serious flood challenges to its downstream residents even though it is much shorter than the Shinano. The smaller Nakanokuchi River, south of Niigata and west of the Shinano, floods with considerable fury, even today.

Perhaps a more dramatic way to appreciate the action of flooding over time is to look at the geological composition of the Echigo and central Japan region. The black areas in Figure 2.4 mark locations of sedimentary deposits and indicate that the region was primarily

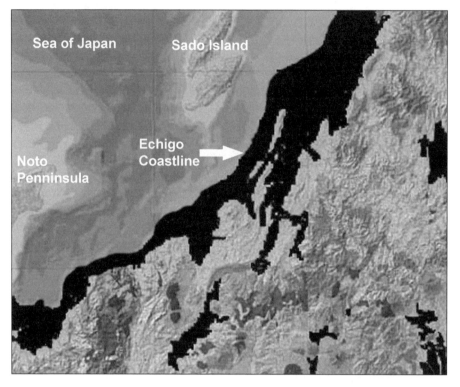

**Figure 2.4**   Echigo Region: Historical Geography

formed through soil deposition. These sedimentary deposits occurred in relatively recent times (the Halocene era and later, in the past 11,000 years). While historical geologists typically are not concerned with developments in historical times, a boring survey for a collaborative research project in Niigata in 2002 indicated that even four hundred to eight hundred years ago much of the area running along the modern Shinano south of Niigata was still ocean. Much of what we today see as part of central Niigata Prefecture's land area was actually formed during historical times by river deposition of soils carried from upland Echigo and the Japan Alps.

The process of filling in the Echigo Plain and other lowland areas continued during the early modern era and into the twentieth century. While much of the Echigo Plain was indeed filled in by the twentieth century, through the nineteenth century there remained many swampy areas and lagoons south and west of modern Niigata City as well as on the smaller plains close to the coastal regions farther to the southwest.

While efforts at drainage were persistent from at least the eighteenth century, real progress on that score had to await efforts based on modern cement, reinforced concrete, and earth-moving equipment that were only widely employed as high-speed economic development took off in the 1960s. In other words, in early modern times, conditions were considerably less stable than today.

Quite apart from flooding, Echigo abuts the *fossa magna:* it rests uncomfortably on this zone and experiences many earthquakes, but also many landslides. Most of the dark gray areas in Figure 2.4 represent debris and various forms of volcanic materials of an age similar to the Echigo Plain's sedimentary deposits or somewhat older. These soils make for many unstable slopes. In the latter half of the twentieth century, Niigata experienced more than 4,500 slope failures. Earthquakes and human construction of highways, rail lines, and the like made their contribution to the twentieth-century experience, but landslides were so frequent, especially in the region of modern Myōkō City (inland of Jōetsu), that special religious sites were established in Heian times in an effort to limit them.

Despite geographic instability and Tokugawa commentary on susceptibility to floods, it would be a mistake to think of Echigo as exceptionally vulnerable to natural disasters or ecologically vulnerable. Among estimates that compare Echigo to other parts of Japan in regard to the natural conditions within which agriculture operated, one indicates that Echigo rates low on the occurrence of major natural disasters; another indicates that it ranked medium high on measures of ecological vulnerability.[39]

## Implications for Japanese Agriculture

The interaction of natural forces and administrative structures had significant impacts on the nature of Japanese agriculture. While rainfall and topography combine to create the flood risk noted above, even without flooding, the annual pattern of rainfall creates variation in water supply more extreme than is common in many parts of the world. The high water volume typically is several hundred times low water volume for Japan's major rivers (the Kitakami, Abukuma, Mogami, Yoshino, Chikugo, and Tone rivers), but varies only from eight to seventy-five times for rivers such as the Seine, Rhine, Thames, Nile, and Missouri. (Niigata's Shinano River is "tame" at a high volume that is sixty-four times low volume.)[40] This variation provides an index of the

variability of water for agriculture, a condition with particularly important implications for production of rice: yields are sensitive to variation
from ideal water supply and quality of water (for example, even when
flooded, paddy water must be refreshed and circulate to prevent afflic-
tions fostered by stagnant water).

To address the timing of water flow, villagers capitalized on Japan's
many hills and mountains to construct thousands of ponds (*tameike*) to
hold irrigation water for later release. Given the small size and widely
scattered nature of a farmer's fields, creation of such reservoirs was
typically a community project, constructed and maintained with labor
contributions from all cultivators. In addition, they created irrigation
channels by directly tapping local streams, also typically an effort of
sufficient scale to require cooperation among villagers.[41]

Yet storing upstream runoff and tapping streams for irrigation water had impacts downstream, and several approaches developed to deal
with competing demands on local water supplies. On the one hand,
villages worked in collaboration to develop and maintain district irrigation systems and to allocate access to water, assuring cultivators of a
fair claim on each season's water supply.[42] In addition, domain authorities invested in large-scale irrigation projects, reinforced and extended
the reach of district water control groups, and adjudicated numerous
disputes over water rights, aiming to prevent upstream activities from
damaging downstream agriculture but still foster expansion of the arable that provided their most important revenue stream. A similar array of activities was associated with flood prevention measures. Villagers constructed small-scale projects—dikes, drainage channels, and
dams—with some district collaboration for larger but still modest projects and domains taking leadership in broader regional projects. Regardless of the source of initiative—village, district, or domain—labor
was typically provided by the villagers of affected areas.

Given the large number of small domains and the scattered rather
than contiguous holdings of the shogun, control of an entire river system, or even its main channel, was typically fragmented. This meant
that coordinated management of Japan's larger streams was impossible for most of the era. Although the shogunate made efforts in the
eighteenth and nineteenth centuries to redress this situation in central
Japan, fully coordinated efforts came only well after the creation of a
nationwide, centralized administration in 1868.

Regardless of coordination throughout a drainage basin, supply
of water was critical not only because plants need water, but because

irrigation water was a significant source of nutrients for rice. All of the nutrients dissolved in water from decaying organic matter meant that soil degradation was less of an issue for rice than for crops such as cotton. While these comments apply to ordinary paddy, rice seedbeds were heavily fertilized with night soil, green manure, and, increasingly from the late seventeenth century, commercial fertilizers such as dried fish cakes. (Even widely produced dry-field staples did not drain soil nutrients to the degree that some cash crops like cotton or tobacco did, and a combination of crop rotations to restore nutrients and use of night soil, green manure, and so forth, were adequate to not only maintain but boost yields over the two and a half centuries of Tokugawa rule. Beginning in the late seventeenth century, the circulation of agricultural treatises communicated the results of keen observation and experimentation with crops throughout the country.[43] These activities were largely at the discretion of individual cultivators and not dependent on community collaboration.)

The preceding discussion suggests the need to think of incentives to invest in the land as falling into two broad areas, those requiring coordinated action at the village, district, and even domain level and those solely focused on the individual cultivator. Irrigation, flood control, and drainage projects were simply beyond the capacity of individual farm enterprises, even large ones. So, too, were many land reclamation projects. In these realms, on the one hand, incentives to act collectively, in cooperation with other villagers and other villages or the domain, were critical. On the other hand, individual effort was essential to maintain the terminus of irrigation networks, assure impermeable paddy pans so that water did not drain from fields, choose seeds and what to plant, and invest labor in plant spacing, weeding, insect removal, and the like. To improve agriculture, incentives needed to focus on both individual and coordinated activities.

## Conclusion

In this chapter I have noted Japanese scholarly emphasis on the instability of the natural environment as a critical factor stimulating the use of joint landownership of arable land. This emphasis suggests local, village initiative in the creation and management of these systems. Domains also became involved in structuring these land rights: there was no national administrative structure that could enforce nationally uniform policies. Indeed, even within a domain there could be variation from village to village regarding the structure of land rights. Both considerations pose formidable challenges for generalizing about land rights in early modern Japan. Echigo, the subject of more detailed inquiry below, also lacked any domain involvement in joint ownership. It followed trajectories of economic development similar to many parts of Japan even though it lacked a single, dominant urban area to serve as a strong market core. The overview above noted in particular that Echigo farmers evinced the same interest in increasing crop yields and exploring commercial crop production that other Japanese regions manifested.

The chapter has also introduced some of the challenges to agriculture and human settlement that Japan's geography presents, challenges that the topographic maps we typically see mask in significant degree. Japan's mountainous topography combined with its abundant rainfall displays a widespread propensity for flooding and landslides, a function in large part of its geological youth: many of its lowland areas formed in just the past several millennia and well into historical times. Erosional forces played a prominent role in plains formation in the past and continue to do so today. While Echigo shows somewhat greater propensity for some natural hazards and ecological instability, it is not qualitatively different from many other parts of Japan.

The preceding discussions raised a significant question regarding the use of joint ownership of arable land: to what degree is the presence of such tenurial regimes clearly associated with susceptibility to flood, landslide, and other sources of change in soil fertility? The Echigo and Kanto plains represent similar geographic circumstances, both linked to a major river, both subject to much erosion activity, yet only the former witnesses widespread use of *warichi*. Are there cases of joint ownership that are clearly divorced from unstable condition of the land? To what extent was joint ownership of arable practiced throughout Japan, and what kinds of variation exist among these rights, over space (as a function of administrative and geographic diversity) and time (a function of changing socioeconomic circumstances)?

The preceding discussion raises other significant issues. If standard economic reasoning is applied, regions practicing joint ownership should not exhibit much interest in improving the land or its output, yet Echigo villagers showed interest in such endeavors similar to other parts of Japan. Are the economists wrong, or do joint ownership regimes somehow make provision for incentives to improve the land? By implication, it is worthwhile to inquire further about the nature of the shareholder-land relationship. Is there evidence that famers' motivation for economic self-improvement was dampened? The emphasis in some explanations of *warichi* on fair distribution raises the question of just how equitably the system operated. Could the redistribution process be manipulated by the village elite for their own benefit? In an effort to answer these questions, it is necessary to consider carefully how incentives and disincentives affected collaborative investments as well as individual investments in land.

The nature of Japan's geography and the administrative context under which early modern Japanese villages functioned create considerable challenges to any effort to address the issues just noted. Precise comparison of geographic conditions in different parts of Japan is difficult and direct comparison of the natural circumstances of *warichi* and non-*warichi* villages impossible. Add to this the administrative diversity of Japan, and the research challenges are compounded. The following chapter discusses the sources I have employed and the methodologies I have harnessed in attempting to "solve" such problems.

# 3

## Data and Methodologies

The discussion in Chapter 2 has a variety of implications for the kinds of data available and the methodological approaches necessary to probe the extent and practice of joint ownership in early modern Japan. To understand the practice of *warichi* well, one must study local practices in widely scattered parts of Japan. Such an approach makes heavy demands—the necessity to understand many local developments in significantly different circumstances. Although I focus most intently on Echigo, I take a multilayered approach to this subject and look at practices in several regions at both domain and village levels and endeavor to place it in a broad, national context. Each level of analysis—national, regional, local—requires different methods and employs different kinds of data. Historians are accustomed to discussions of historical sources, but discussion of methodologies is less common; social science readers are used to discussions of methodologies, but discussion of data usually focuses on circumstances considerably different from those confronted here: they are more likely to work with survey data generated from interviews and questionnaires or statistical data compiled nationwide by a modern government. Given the somewhat unusual circumstances that this work presents to both audiences and the importance of understanding both the characteristics of the evidentiary base and the methodologies, I provide an overview here.

Documentary evidence frequently does not speak directly to the questions of modern historians, a situation that leads me to employ a combination of approaches to adduce answers. Some of these approaches, the use of contemporary documentary evidence and surveys of secondary literature for example, are broadly familiar to historians and social scientists alike. Others, like the use of geographical information systems software, have been employed in historical studies only recently and are not even widely employed in the social science disciplines outside geography; these approaches require a bit more elaboration.[1]

## Multilevel Approach

I take a trilevel approach to the subject. At the highest level, and in order to assess the extent of *warichi*, one needs a sense of its distribution over the three main islands of early modern Japan: Honshu, Shikoku, and Kyushu (Hokkaido was not yet well encompassed by Japanese suzerainty). For most of the period of concern, there was no central government with the desire or capacity to compile Japanwide surveys of land tenure systems. This means that there are no national data on land tenure arrangements during the Tokugawa era: estimates of the geographic scope of joint ownership practices and estimates of its relative importance within Japan during the early modern era must be assembled from a survey of readily available sources. So rather than the historian's privileged primary sources, contemporaneous with the era under study, I surveyed secondary scholarly literature. That effort has been aided by two far-reaching, prior efforts, a bibliographic survey of prewar studies of joint ownership practices by renowned rural historian Furushima Toshio and Aono Shunsui's major postwar cross-regional study.[2] I supplement these works with material from more recent journal publications, my own on site surveys and interviews with local practitioners. This survey provides the first attempt to offer an estimate of the proportion of arable land in early modern Japan that was, at one time or another, under joint ownership regimes.

The lack of systematic national data also means that surveys of secondary literature and interviews play a particularly important role in creation of a typology of joint corporate landownership. This review is sufficient to establish baselines for assessing the variety of *warichi* practices. These categories, too, offer the first comprehensive effort to conceptualize joint ownership systems in Japan.

Second, I employ daimyo domain-level data to examine two key examples of domain-managed *warichi*. Even at this level, there are often significant lacunae in primary source materials. While domain ordinances may be reasonably complete, serial data on actual practice within the domain is conspicuously limited. Some efforts that call for diachronic analyses therefore draw on documents that represent an incomplete record. Among these sources are notebooks that record the outcome of reallocation of access to land, occasional settlements of disputes over redistribution, listing of basic procedures to be followed during a reallocation, and similar documents. I supplement these contemporary records with data from secondary sources, especially where I have not had access to documents because of loss or limits of time. For the most part, domain-level analysis is limited to two domains, Satsuma in Kyushu and Kaga in the Hokuriku district on the Japan Sea opposite modern Tokyo.

The choice of Satsuma and Kaga for study is partly fortuitous. My discovery of the joint ownership phenomenon was the outcome of my early work on Kaga domain, a logical choice since it also turned out to be one of the better-documented cases. The choice of Satsuma came by default. Early in the process of looking for a regional case on which to base my study of joint ownership of land, I spent the fall of 1991 traveling to all locations in Japan where I knew it had been practiced, from Tohoku to Okinawa, consulting with local historians about the availability of both published and manuscript data. My survey made it clear that even where I could identify the radically wealth-redistributing forms of the practice, documentation was very poor. In one case, Okinawa, documents had largely been destroyed by the U.S. assault and related damage during World War II. For the study of Satsuma domain's famous *kadowari* variant of joint ownership, although there existed limited data, eminent historian Ono Takeo had conducted his own surveys during the prewar years, through both interviews and documentary research, and published the results. This provided sufficient data to prepare an overview of this particularly interesting form of joint ownership.

My 1991 excursions also led me to the town of Kumayama, in Okayama Prefecture, about thirty minutes north of Okayama City. My purpose was to visit with Hiroshima University emeritus professor of geography Ishida Hiroshi, who was then overseeing the compilation of his hometown's history. By the time I arrived, I found that my preliminary inquiries had led him to explore why certain parts of the town bore *wari* as part of their name, written with the same Chinese character as the *wari* in *warichi*. As he talked with friends and other residents, he was able to confirm several instances in which *warichi* had been practiced up into the 1970s. This serendipitous discovery introduced me to a new type of joint ownership, one that I could document through interviews with people who had actually managed lands under joint ownership.[3]

At the conclusion of my investigations, I chose Echigo Province as the focus for my study. The practice was typically village-based, and Echigo offered promise for exploring the relationship between the natural settings of villages and joint ownership practices. In addition, villagers in Echigo employed the most common form of joint ownership of arable land. Printed and manuscript documents collections with which to explore village practices and domain administrative contexts were more plentiful than for other regions of Japan. Local historians' surveys of *warichi* helped the compilation of a provincewide sample for analysis. Finally, the province encompassed a wide variety of geographic circumstances, from landslide-prone hill and upland areas in its southern districts to the narrow mountain valleys and finally the flat floodplains of the lower Shinano River. Based on my efforts to survey the national scope of joint ownership, I believe Echigo *warichi* to be generally typical of village-based joint ownership practice as most widely found in early modern Japan.

Although some of the following analysis takes a provincewide approach, I primarily exploit village-level operation of *warichi* in Echigo, the third level of my analysis. Village documents provide the foundation for looking at the details of how joint ownership functioned. This analysis generally makes use of the traditional tools of historical research and is based on analysis of both printed and manuscript documentary materials.

## Documentary Characteristics and Limitations

The reasons for my choice of Echigo as a case study notwithstanding, even there limited documentation remains with which to explore actual operation of *warichi*. Where daimyo administration made an effort to regulate some or all of the process, we have domain ordinances, but such ordinances were always subject to local interpretation, and they often allowed for considerable local variation in practice. Scattered district and county officials' decisions in redistribution disputes that came before them also remain.

Nonetheless, readers should note that the large degree of self-governance by Japanese villages meant that most routine matters, and even a large number of internal disputes, were settled solely within the village, often without leaving written records. Unlike early modern European traditions, in which courts and the provision of justice were seen to be public goods, in Japan, adjudication of disputes continued as a largely private prerogative of the ruling classes, often not separate from other administrative functions including taxation. While in urban contexts legal specialists developed to advise those engaged in lawsuits, most rural residents could only appeal to a village group headman (e.g., a *tomura, ōjoya,* or *ōkimoiri*) or a low-level domain intendant *(daikan)* whose primary function lay in tax collection and assessment and other similar duties. Bringing such officials into village business might result in settlement of a dispute, but they also might use the opportunity to reconsider and to raise land taxes or in other ways create additional difficulties for a hamlet.

Thus, while court records are often very useful for European, American, or even modern Japanese social historians, in the case of rural administration in early modern Japan, there were strong disincentives to involve supravillage authorities, and most disputes over *warichi* were handled as a strictly internal village affair.[4] Domain administrators only rarely became involved and typically only in conflicts between villages, where access to supravillage authorities was unavoidable if the villages involved could not reach agreement on their own.[5]

Consequently, court records that deal with *warichi* are rare for the Echigo region. The examples I have found all involved a variation on the theme of resolving diversity disputes: the complaints were filed by someone from outside the village who came to manage land within it as a result of marriage or other similar familial connection. These circumstances also meant that cases presented to district officials dealt with a limited range of issues (often, whether nonresidents could participate or whether a village traditionally followed redistributive practices or not). So while there are relatively rich records for disputes over water management, which by its very nature involved diverse village and regional jurisdictions, records of disputes over joint landownership in the core areas of study are countable on half a hand.

In addition to court records, historians studying *warichi* commonly analyze a variety of village documents. Village regulations that described the way in which a particular village implemented redistribution constitute one major source. Such documents are useful, but only up to a point. Commonly we find only one or two such documents for a given village, typically for the late eighteenth and nineteenth centuries, and therefore evidence does not reveal historical trends in practice over time. The shortage of such documents is likely the result of several factors. One significant factor is the lack of any direct connection to daimyo, *hatamoto*, or shogunal administration and the high degree of internal village autonomy just noted. Where documents did not document the village's relationship to domain administration, villagers lacked compelling reasons to save copies to verify past dealings with *han* authorities. In the context of village administration, there was little need to retain copies of any documents except those related to the most recent instance or two of a redistribution. These conditions hinder analysis of the ways villagers changed the process of redistributing land over time.

The notebooks that reveal the outcomes of a *warichi* redistribution constitute another manuscript source that historians have exploited. They provide essential data for understanding how these systems worked in practice. Analyzed seriatim, they require detailed comparison, one iteration to another, to yield insights about change over time. Yet manuscript copies of these notebooks are few in number, commonly treating only one or two iterations for a village. The exceptional Satō Family Archive provides the core of my analysis of village practice in the chapters that follow. This collection contains an extended series of notebooks that reveal how the village managed its joint ownership practices from the early eighteenth to the mid-nineteenth century.[6] In addition, other manuscript evidence cited below comes from documentation of the repeated efforts to reclaim land from the waters of the Shinano River.

I supplement these more traditional approaches in a variety of ways.

## Interviews

The practice of joint landownership continued after the implementation of modern forms of land taxation in the 1870s. Initially, the joint landownership survivals were relatively numerous, but instances became increasingly rare with the passage of time. Despite this trend, some communities continued the practice well after World War II.

Post–World War II survivals have provided an opportunity to interview people who had direct, personal experience with joint landownership. This was true for a handful of people in modern Tokamachi City and Nagaoka City in central Niigata Prefecture as well as one individual in Kumayama, now a part of Akaishi City in Okayama Prefecture. In some instances, the land that was so held comprised small segments of the traditional late Tokugawa village. In other instances, parts of Nagaoka and Tokamachi, for example, the last areas subject to joint ownership were the fragmentary remnants of a practice formerly applied much more broadly in the community. Practices of management and rotation of jointly owned land in the twentieth century were not necessarily identical in mechanism or function to earlier counterparts. Several cases of joint ownership practice clearly reflect a transformation over time. Nonetheless, the opportunity to interview practitioners—the result of the adaptability of the practitioners of this form of ownership as well as their sense of commitment to the practice—provided useful insights that supplemented documentary analysis.[7]

## Geographic Information Systems Technology

As remarkable as the use of oral history is for a Tokugawa era project, perhaps the most innovative methodology is a relatively new tool in historical research, geographic information systems (GIS) technology, employed with digital elevation and other data in order to assess the degree to which environmental factors were reflected in the redistribution practices of jointly owned lands. Using terrain and other geographic data from the Japanese government (the National Land Agency) and tabular data I developed during my research, I used ESRI's ArcGIS software package to create all of the analytical thematic maps and a number of other maps below.[8] (Many of the data sets are now freely downloadable from web sites associated with the Geographical Survey Institute of Japan [www.gsi.go.jp] and on CDs marketed by the Japan Map Center [Tokyo].)

Given the role of environmental factors influencing agricultural productivity and land under the plow, the use of GIS permits comparison of multiple village environments and their relationship to joint landownership practices. Comparison is most effective within small areas, helping to hold soil type, climate, and other factors constant for the villages under comparison. However, this kind of analysis sheds some light on broader regional comparisons as well. This type of systematic regional assessment has not been attempted in any previous study.

GIS processes and data may not be familiar to readers, so some explanation is in order. It is easiest to think of GIS software as a way to generate digital maps on which to display data that correspond to a particular latitude and longitude. Like their paper counterparts, these maps show the location of a site in relation to the surrounding environments, whether natural or human, within the context of a specific model of the earth's surface. Rather than plotting data by hand (e.g., using colored push pins on a map to show the distribution of evangelical Christian and Catholic churches, synagogues, and mosques in New York City), one can enter latitude and longitude coordinates into a data table and use them to link those data to locations for display on a digital map. The data can be for a point (e.g., the location of a school), a line (e.g., a stream), or a polygon (e.g., the boundaries of a county).

Despite the capabilities of the technology, any attempt at broad cross-regional and national comparisons of natural environments and their impacts confronts confounding circumstances. Japan's three major islands exhibit a wide variety of soils, slope, vegetation, and climate. Ideally, to compare different features of local joint ownership practice in one environment with that in another, one needs to be able to identify and measure these diverse environmental factors. However, at present, there is no way to create an effective index that allows their reasonably precise measurement. Circumstances vary so much that just because there are nationally or regionally widespread natural disasters, for example, it does not mean that they affected specific villages or districts, or that in years of nationwide natural quietude, local communities did not suffer significantly from floods or landslides. All of this complicates comparison of the degree of natural hazard risk in two locations, one practicing joint ownership and one not, and assessing the degree to which joint ownership is functioning as a device to minimize the impacts of floods, landslides, or microclimatic and microgeographic variation.

Thus, while a broad regional comparison, as noted briefly in Chapter 2, can be used to raise the question of the degree to which geographic differences can explain the presence or absence of joint corporate control over arable land, they cannot provide answers. I therefore examine land redistribution and variations in its practice in a much smaller region. This approach offers a better opportunity to limit or eliminate variations such as soil type and climate patterns. It is with this purpose in mind that I examine *warichi* in Echigo Province later in this book.

More specifically, my solution is to locate village hamlet latitude and longitude and then use that information in digital maps to assess the influence of the natural environment on the village practice of joint landownership; I compare villages that are in very close proximity to each other, typically within less than a mile of each other. This minimizes the variation in natural conditions—climate, soil type, and the like—as well as in economic conditions and allows us to see how village communities responded to similar natural and economic stimuli. In addition, such an approach can be used for comparison within a somewhat broader region to further question standard arguments for the close link between the natural environment and this practice.

One advantage GIS provides is the flexibility it permits. For example, the digital elevation model (DEM) data on which my *warichi* data are displayed permit a resolution of elevation detail far greater than is possible with standard paper maps. Unlike with paper copies, the researcher can flexibly change contour interval, area represented, and other map characteristics. One can create and manipulate maps in other very useful ways.

For example, it is possible to scan a paper map, import it into a GIS environment, and assign latitude and longitude coordinates to it. Data can be added to this map just as with a map created from digital elevation models. This I do below with modern maps of land classification and natural hazards. Further, it is possible to use such "georeferenced" maps in GIS to identify coordinates for villages on the map and then enter them into a data file. Below I adopt this approach so that some Edo era villages can be displayed on a digital map.

The biggest initial problem in using GIS for historical work lies in assigning coordinates to village locations. While Tokugawa villages did not produce location data useful in a GIS database, remnants of their activities, including settlement clusters, have been transmitted through the ages and can be rendered usable in digital data projects. With hamlet coordinates I have been able to link a village to its natural surroundings and to other data on *warichi* to explore the relationship between villages, their natural environments, and joint ownership at a local and regional scale in ways not possible before.

## Determining Latitude and Longitude of Early Modern Villages

Identifying the latitude and longitude for early modern hamlets presents a challenge. Japan's first modern national cartographic surveys were not undertaken until the late nineteenth century. Famous Japanese cartographers such as Inō Tadataka experimented with modern techniques, and Japan's modernizing Meiji administration was relatively early in testing the use of modern cartography. Their first modern surveys focused on the major metropolitan areas of Tokyo and Kobe-Osaka-Kyoto in the 1880s. More than a decade later, Japan undertook a modern, nationwide 1:50,000 topographic survey, extending the use of modern cartographic techniques to less densely populated regions.

In the interim, Japan had witnessed substantial administrative reorganization at all levels. It began with the abolition of old semifeudal daimyo domains and the creation of prefectures in the 1870s, a process that frequently ignored the old administrative boundaries. In addition, cities, towns, and villages were reshaped through amalgamation.[9]

Far more dramatic than the formation of prefectures was the new government's abolition of tens of thousands of premodern villages as distinct administrative units, an effort to lower the costs of administration and to place the government on a more stable financial footing. By the mid-Meiji, the number of rural hamlets had been reduced to one-sixth of their number at the time of the Meiji Restoration in 1868. In other words, most late Tokugawa villages lost their administrative function when they were amalgamated into larger villages; they were never mapped as governing units by modern cartographic surveys.

It goes without saying that any research that explores the interrelationship between a Tokugawa era community and its natural surroundings requires clear identification of hamlet location—either by center point or by boundary. While village boundaries certainly changed during the seventeenth to mid-nineteenth centuries, most such changes involved the breaking off of part of a village to form a new one, not amalgamation of villages, as has been the case since 1868. As a result, the villages present in the mid-nineteenth century and the first years of the Meiji era represent the largest number of villages to have existed in Japan's history.

For some purposes, recovery of lost village boundaries is essential; however, for other research purposes, simply locating some reasonable center point—the center of a village's residential cluster—is adequate. That is the approach I have followed. Premodern Japanese villages generally clustered houses together, with fields distributed around this core. This pattern continued to prevail well into the twentieth century, aiding identification of village center point latitude and longitude.[10]

For a small study region, one can determine village location through site visits, interviews with local informants, and archival research, but there are growing limitations to this approach. One problem is the mortality of informants familiar with old neighborhoods. Increased urbanization and suburban development compound the problem, obliterating distinctions among old hamlets. In some cases, these processes dramatically transformed the land and destroyed physical boundaries. For example, one can now travel around many parts of modern Niigata City in a car where, in 1960, travel would have required a boat.

When a study requires data across regions, the use of local informants is impractical. The time and effort required for site visits, research, and interviews would be prohibitive. Consequently, some efficient means of recovering early Meiji location for significant numbers of villages is essential.

To deal with the problem of identifying hamlet latitude and longitude, I have followed two approaches. The first employed residential hamlet location data for rural communities that appear in the Japanese government's surveys of agricultural hamlets. The national Ministry of Land, Infrastructure, and Transport (formerly the National Land Agency) has made increasing amounts of data available at a reasonable cost. Among these data are the locations for features on its 1:25,000 surveys. These features include many customary names for what people today might call a neighborhood (now often referred to as an *aza, ko-aza,* or *ōaza*), but which often represent old Tokugawa villages.

Such a correspondence between modern and nineteenth-century data could not be presumed since the current government's administrative interest lies in contemporary hamlets. No geographer I consulted was able to explain how the government's modern location data were determined. I found no explanation for what data points were employed.

To assess the utility of this Japanese government data, I explored hamlet location with a local informant, using a global positioning system (GPS) receiver to identify latitude and longitude of the residential clusters in one rural part of Niigata Prefecture. The choice of region for this experiment was based on a convergence of beneficial circumstances. The region was familiar to me from previous research; I also knew local historians who were familiar with changes in the area based on their own research for the multivolume local history as well as lifetime residence in the area.[11] The area chosen is near modern Yoshika-wa-machi (Plate 1), now a part of Jōetsu City.

As a preliminary test, rural areas such as this represent something of a best-case scenario. Rural areas are least likely to have experienced the deforming destruction of war or the construction of metropolitan complexes and their associated suburbs. More densely populated regions present greater challenges. Nonetheless, since much of Japan remains rural, the results of this test promise to be broadly applicable.

Since the details of my test procedure have been published elsewhere, I simply introduce the outcome here.[12] Figure 3.1 displays data for a selection of the four dozen villages I surveyed in my test. The government data are represented by circles, and the GPS observation data are represented by Xs. Note that my GPS observations and the Japanese government data coincide. Rounding in data coordinates masks some variation in the GPS and government data, but the difference, even in the worst two cases, was less than 10 to 20 meters, certainly close enough to verify the utility of the Japanese government data in the context of my research objectives.[13]

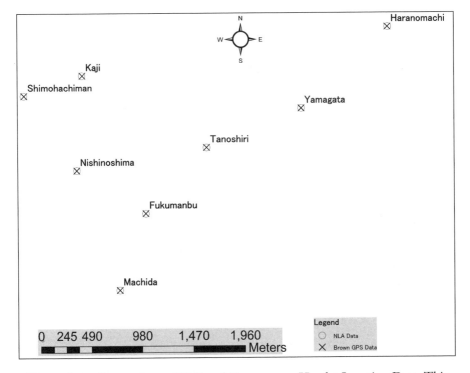

**Figure 3.1** Comparison of GPS and Government Hamlet Location Data. This map accurately places hamlet latitude and longitude locations for both data sets, and distances are accurately displayed. However, to focus attention on the interrelation between the two data sets, these data have not been overlaid on a topographic map.

For purposes of this study, no effort was made to identify village boundaries. First, any boundaries initially identified in the Meiji era would have had a very different meaning before that era than we commonly ascribe to them today. Overlords were primarily interested in arable lands, and that is what they surveyed most closely. They also identified forest and upland resources necessary to build their own fortifications and to use for hawking and similar paramilitary exercises. They paid far less attention to marginal lands that might be treated as a commons. Villagers, by contrast, were concerned with the delineation of the realm from which they had a right to draw resources.

Village conceptions of boundaries were more fluid than we typically presume today. To take just one example of commons, access might well be shared with one or more villages. When access to commons is portrayed, maps of a Tokugawa era village might include all of the shared commons, and a map for another participating village might also include that same shared commons. Two entities would appear to be claiming the same territory—a phenomenon that might suggest to twenty-first-century map readers that this was contested territory, when in fact that was not the case at all. In addition, maps of these same two communities compiled for different purposes, for example, to show irrigation networks, might well depict very different boundaries than those that had included the commons, even if made at the same time. So Tokugawa maps, like modern maps, were shaped by the functions they were designed to serve, but unlike in modern maps, the concept of an administrative boundary encompassing complete control of resources by one town to the exclusion of a neighboring entity had not yet achieved dominance. Even if we had some means of locating late Tokugawa or early Meiji boundaries, that task would be too much to accomplish for the number of villages surveyed here. Even Japanese scholars have not attempted this on other than a case by case basis.[14] Furthermore, in order to understand the natural environment in which villages are set, boundary data is not meaningful. For this purpose we are interested in natural phenomena that transcend a single village, for example, its location in a drainage basin.

While the Japanese government data can be used to efficiently gather location data for a number of early modern villages, this method did not work for all hamlets for which I had data. To get coordinates for villages not included in the Japanese government data, I employed a supplementary approach. I digitized the earliest of Japan's modern maps, the 1:50,000 topographic maps. Once georeferenced, a click of the mouse identified the latitude and longitude of the place where the cursor touches the map image. Careful perusal of several of these maps allowed identification of the latitude and longitude of almost all early modern hamlets in my data sample but not in the Japanese government data.[15] Between these two approaches I was able to obtain location data for all but one or two villages I wanted to use.

Mapping locations allowed me to situate villages in their natural surroundings and to assess how villagers used *warichi* in relationship to key natural hazard risks. There are more subtle variations in the productivity of land that joint corporate ownership might accommodate, but we have no way to trace them over time and space. Even in the case of flood and landslide risk, their incidence can be affected by human activity. Harvesting of timber, overextension of arable land, construction of dikes, dams, drainage channels, and storage ponds, and many other activities affect the incidence of floods and landslides as well as the water supply for agriculture. Nonetheless, the procedure outlined here, in combination with the examination of cases in very close proximity—to minimize differences associated with climate, denuding of hillsides, and other factors—provides a basis for assessment of the degree to which villages in similar environmental circumstances responded in like manner to threats of natural hazards.

The preceding discussion of data and methods provides some sense of the challenges the data presents, the scope of the study undertaken here, and some of the solutions I have employed to address the challenges. In the next chapter, I start at the national level, first exploring the varieties of joint ownership and the effects of the different mechanisms they use to allocate cultivation rights, presenting examples of each type, making an assessment of the geographic distribution of joint ownership practice and assessing the share of early modern Japan's arable land subject to these regimes.

# 4

## Varieties and Extent of Joint Landownership

The preceding chapters have demonstrated the structural potential for domains and villages to develop multiple arrangements for regulating farmer rights in land. They also hint that geographic conditions alone may not explain the presence or absence of *warichi*. The presence of joint ownership itself is a significant indicator of the variation in landownership types permitted by and introduced under Japan's early modern administrative structure. This chapter presents further evidence for that diversity and explores the major patterns of joint ownership practice.

Just considered on its own, the diversity of practices within *warichi* is fascinating. Why did different regimes develop? What were the differences among them, in the way they functioned and in their effects? Did differently structured land rights result in different incentives to invest in land improvement or to encourage free riding?

Explanations for joint ownership practices and their impacts are not always easy to untangle, but an effort to tease out the effects of these landholding systems helps us understand more about the nature of early modern agricultural practices and the significant ways in which domains and villages functioned as corporate entities. This diversity in practices challenges some of the widespread beliefs about the relationship between specific rights in land and incentives: the belief that joint ownership necessarily entails inefficient, counterproductive incentive structures as well as beliefs about the relationship between joint owner-ship and environmental conditions.

I conclude the chapter with an effort to address several other key questions: To what degree did early modern Japan witness joint land-ownership? Was it simply an outlier, the exception that proves the rule? Or was it more than a fluke, an odd practice restricted to isolated, small communities?

## Varieties of Joint Landownership

The highly dispersed control over rights in land is not just of interest in the context of *warichi*; other variations also existed. Records compiled in the course of preparing for the Meiji Land Tax Reforms of the 1870s show practices that range from a variety of joint ownership regimes to multilayered fragmentation of rights to a single plot. Noted land tax reform scholar Fukushima Masao divided these practices into two broad categories: multilayered land rights and redistributive practices (what I refer to as *warichi*). In the case of multiple landholding rights, two or three individuals each held specific rights in the same piece of land.[1] This kind of practice was akin to a contemporary landowner sell-ing rights to drill for natural gas on his or her land.[2]

My purpose here, however, is to explore the diversity of joint own-ership practices on arable land. Variation raises additional questions about the original functions of *warichi* and its relationship to natural hazard risk. Joint ownership broke the direct tie between a villager's investments in land and the potential rewards of that investment. The manager of the land did not, through his own volition, control which piece of property he managed. Access to a given piece of land was con-trolled by villagewide or domain regulations, implemented through the village, calling for redistribution under specific circumstances and us-ing a variety of mechanisms and principles. These schemes highlight the continued weight of some form of control of arable land by a pub-lic authority or multiple authorities (e.g., village and domain) in early modern Japan.

The diversity explored here raises the question of whether a variety of problems and issues are seen in different types of *warichi* in a similar way and to a similar degree. To what degree do structures that lodge control of arable land in the hands of villagers acting jointly remove in-centives to invest in the land as widespread understanding has it? Is this really the case in all *warichi* regions? Do all varieties of these structures create free rider problems in which some participants can successfully seek and get something for nothing, for example, soil nutrients without having to renew the soil?

In all of the joint ownership regions, substantial nonmarket communal controls mediated a farmer's access to land, yet the forms and extent of communal mediation were not all the same. Why? Why did some domains take a direct interest in promoting nonmarket controls on villager access to land rights? Why did some take a more hands-off approach? What were the consequences of some of the differences in the form and nature of nonmarket mediation of access to land?

This chapter begins to explore whether villagers under various forms of *warichi* somehow managed to avoid negative outcomes. Answers provided by the three case studies presented in this chapter are highly suggestive. The following analyses mark a useful point of departure for resolving these and other issues of joint ownership, concerns to be explored further in later chapters.

To begin, I classify *warichi* into three basic types. This categorization is based on my readings of secondary sources, primary source materials, and discussions with living practitioners of *warichi*. To my knowledge, it incorporates all major varieties of *warichi*. Table 4.1 presents a schematic overview of types of redistribution systems.

The primary criterion for classification is the allocation principle that guides each type, a principle I have tried to capture in the labeling of each variety. These labels were not found in practice. They are my own, and readers should bear in mind that because of the decentralized locus of control, in the field one sees myriad variations in nomenclature and details of practice that, for reasons of space, cannot be elaborated here.[3] The columns in Table 4.1 for each redistribution type are arranged from least common (Type I, equal per family redistribution) to most common (Type III, per share redistribution).

Type I, equal per family redistribution, distributes land of equal absolute yield to each shareholding family in the village. Land area might differ slightly, but the village divided lands into sections they considered of equal overall output. This approach did not redistribute wealth among villagers. No access to land was taken from one family at the expense of another. In examples of this type that I have found to date, only a small fraction of village arable land was subject to redistribution. Consequently, sharing of risks of flood, drought or landslides, or sharing of investment gains applied only to this small portion of a villager's lands.

**Table 4.1** A Typology of Joint Ownership of Arable Lands

| | I. Per Family | II. Per Capita | III. Per Share |
|---|---|---|---|
| Redistribution principle | Same value per family | Based on family age/gender composition | Based on shares/cultivation rights held |
| Redistributes | Some arable | All arable | All arable |
| Redistribution mechanism | Sequential rotation | Lottery | Lottery |
| Locus of control | Village | Domain or village | Domain, village, or both |
| Wealth redistribution effects | None | Explicitly, heavily wealth redistributing among families | Generally modest based on shared investment in land, land lost from use or declining quality |
| Equalizing effects | Yes (per household) | Yes (by designated elements of household composition) | No (with possible exception of joint reclamation projects, depending on valuation of individual investment) |
| Periodicity | Fixed interval | Fixed minimum intervals; more frequently possible | Fixed interval or significant change in endowments, e.g., major flood |
| Purchase/sale | No | No | Yes |

*(continued on next page)*

**Table 4.1** *(continued)*

|  | I. Per Family | II. Per Capita | III. Per Share |
|---|---|---|---|
| Key potential disincentives | No incentives for individual reclamation; possible disincentives to maintain/improve land as redistribution date approaches | No incentives for individual reclamation or land improvement since land reallocated frequently | Greater disincentives for land improvement where redistribution intervals shorter; incentives to maintain or improve land decline as redistribution nears; possible disincentives for individual reclamation |
| Primary effects | Equal sharing of rewards from investment (e.g., reclamation); assure minimal food supply; raise revenue (e.g. shrine) | Assure basic food supply; protect supply of communal and agricultural labor for village and domain | Risk sharing for changes in natural endowments; share rewards from investment (e.g., reclamation) in proportion to inputs |
| Secondary effects | Equalization of taxation on redistributed land; adjustment to changes in arable (land reclamation; loss due to flooding, etc.) | Intravillage equalization of effective average tax rates per unit of productive value; adjustment to changes in arable (land reclamation; loss due to flooding, etc.) | Intravillage equalization of effective average tax rates per unit of productive value, labor retention; encourage reporting of all land for taxation |

In contrast, Type II, per capita redistribution, specifically distributed access to agricultural lands based on changing family composition. The amount of land a family cultivated would vary over time depending on the age and gender of the members of the household. Of the three types of redistribution it is the most radically wealth-redistributing. In principle, all arable under village control was subject to redistribution. Risk sharing existed throughout the village land; incentives for investment operated primarily at the village, not individual, level.

Type III, per share redistribution, allocated land to villagers in proportion to the cultivation rights that each family held. It has some wealth-redistributing effects, but they are more limited than for per capita redistribution. For the most part, the redistribution effects were limited to sharing proportionally (a) loss in village arable land, (b) decline in natural endowments of village land, or (c) increases resulting from land reclamation. The structure of the redistribution process assured that each participating villager received cultivation rights to all types and qualities of land in the village, and no one individual could exercise rights only in one part of the village.

As this brief description suggests, joint ownership exhibited an array of wealth-redistributing effects. It did not just exhibit the wealth-equalizing impact that contemporary scholars often presume based on the widely known cases of land reforms of the twentieth century, for example, in Russia, China, Zimbabwe, and several Latin American countries. Indeed, its wealth-redistributing effects could be very limited or nonexistent.

Consistent with earlier discussions of the dispersed responsibility for defining rights in land, control over redistribution might reside with domain administrators, village leadership, or a combination of the two. The effort of different domains and villages to work out their own arrangements is also evident in the noteworthy variation in the proportion of arable land in a village subject to redistribution, from all arable land to what could be just a tiny fraction of the land in a village. The limited area of land involved in Type I redistribution permitted a simple periodic sequential rotation of access to the fields to which cultivation rights were assigned. For the other types, more complex allocation procedures, typically based on a lottery, were implemented.

Note, too, that the outcomes of redistributions varied. Type I guaranteed families access to a limited resource that constituted a fraction of village arable land, provided for shared responsibility for a community project such as shrine maintenance, or shared the costs of reclaiming land in a jointly developed project. The first two functions have not been found in any case of Type II or Type III redistribution I have encountered. The third, provision for sharing the benefits of a joint investment in extension of arable land, has also been an outcome of Type III redistribution. Data for Type II do not speak directly to this point, but certainly village investments in expanding arable could have been accommodated within its operating principles.

The potential to ameliorate the impact of changes in land endowment, a significant outcome for Type III, is present but not prominent for the other two types. In different ways, both Type II and Type III maximize the possibility that a family can continue as a successful farm enterprise even under adverse circumstances caused by floods or other natural hazards, Type II by allocating fields based on family needs as determined by each family's composition and indirectly accommodating endowment changes, Type III by sharing the costs of floods and similar changes in land endowments through proportional reallocation. In Type I redistribution this element is limited to a fraction of village land if it operated at all.

To the degree that family farmsteads were retained in the village, labor for communal and domain projects was also fostered by implementation of Type II and III redistribution mechanisms. This was accomplished by sharing losses of land from cultivation but also by promoting fair taxation. All three systems promoted fair taxation based on uniform rates per unit of productive value (as will be the case throughout this book, Japanese emphasized allocating lands with comparable productive value rather than equal units of land area).[4] That said, since Type I applied only to a small fraction of village lands, the tax consequences of redistribution were very minor.

The equal access to land per family in Type I and based on family composition in Type II precluded the sale and purchase of land and restricted inequality of cultivation rights. However, under Type III, rights to cultivate land could be freely bought and sold. This permitted the growth of considerable landholding inequality among families.

The growth of inequality in rights to cultivate land raises the big question of why large landholders continued to tolerate redistribution systems, at least in the case of Type III. In other, more common forms of fee simple ownership, one would expect that a wealthy landowner would use that wealth to acquire all of the best land in the village and leave poorer land or land more vulnerable to floods or landslides to the poorer farmers. That was not possible under Type III systems. This is a question to which I will return later, but for the moment, it is sufficient to stress that Type III did permit and did witness considerable inequality of wealth.

The fact that land is reallocated to cultivators on an intermittent or periodic basis also raises the possibility in all of these regimes that cultivators might either reduce investments in land or not make any at all since the benefits of such improvements could be taken from them. Although all of these systems have such a risk, it is not the same for each. First, the interval between redistributions was likely critical. For Type II, where allocations are attuned to changes in family size and/ or composition, intervals were generally short and presented limited opportunities to capture the rewards of individual investments in land quality when nutrient-draining varieties of crops were grown or other special circumstances prevailed. For the other two types, there is no general tendency toward either short or long intervals that I can identify, and much evidence indicates great variation in interval. While I have found only a smattering of cases for Type I, the interval has tended to be about every five years, but there is nothing inherent in the structure of the redistributions to have prevented either longer or shorter intervals. For Type III, observed intervals range from annual redistributions to two decades or more (see discussion of Kaga below). In addition, a number of Type III regimes operated without a fixed interval, and redistributions might take place only once every four to eight decades. In these latter cases, residents might have had a sense about their likely redistribution chances based on past practice but no certain knowledge of future redistributions. Again, if the expectation were that redistributions occur frequently, then the potential disincentive for individuals to invest would be higher than if the expectation were that redistributions occurred infrequently. This, too, is a question to which I will return later in this book.

To this point I have discussed villager interests in joint ownership, but since domains sometimes became actively involved in regulating tenure regimes and encouraging redistribution, it is necessary to address their interests as well. Domains sometimes promoted Type II and III redistributions that affected all arable lands in a village. To some degree, domain and villager interests overlapped. Promotion of fair taxation is a case in point. To the degree that taxes were fairly distributed among villagers, it was less likely that a farm enterprise would fail and that a family would continue to be productive farmers, contributing revenue to domain coffers. But in addition, it meant that these families continued to owe labor dues for domain public works projects. Both Type II and Type III promoted these ends as either primary or secondary outcomes. Finally, for Type III, there is some evidence that domains viewed redistributions as a mechanism for ensuring that villagers registered all taxable lands with the domain.

## Three Case Studies

Discussion of specific cases will help to clarify and emphasize the variety in redistributional systems and their impacts that have been schematically presented in Table 4-1.

*Per Family Redistribution (Type I) in Kumayama.* Although Type I represents a small fraction of the land on which *warichi* was implemented, even a brief discussion illuminates some creative uses of joint ownership. Kumayama Town, now part of Akaiwa City in Okayama Prefecture, actually presents two examples. One concerns rotating cultivation responsibility for fields that supported a local Shinto shrine; the other provided village families access to a scarce resource, dry fields, critical to residents' subsistence farming before the widespread advent of neighborhood grocery stores and supermarkets.

A small section of modern Kumayama was owned by a local shrine and was rotated in a regular sequence among members of the shrine association. Cultivation of this land provided grain and other offerings for shrine activities. Some of the produce raised on it was sold to finance the shrine. If there was any surplus after providing for shrine activities, the cultivator retained it. Responsibility rotated annually.

The use of redistribution in support of a shrine or other community activity is interesting but very rare; somewhat more common is the use of *warichi* to allocate a limited resource among village families. The modern *ōaza* of Seiriki employed *warichi* during the nineteenth and twentieth centuries in the Shimizu section to provide dry fields for households in the village. Redistribution was applied only to this area and assured that all families had land on which to grow vegetables, a major source of nutrition for each family. In the nineteenth century, at the end of the Tokugawa era, this community constituted an independent village.[5] Unlike many villages, it possessed only paddy, so dry field for production of vegetables, while possible on the earthen paddy ridges, was simply inadequate. By dividing the limited dry field available into segments of equal per hectare productivity and rotating cultivation rights among the village families in a fixed order periodically, all of the families in the village were assured access to sufficient lands to produce their vegetables. This arrangement functioned as a closed system, and immigrant families were not permitted to participate.[6]

Since there are no records that document the origin of this custom, it is difficult to ascertain why each family, regardless of wealth, size, age, and sex composition, was allocated access to equal shares of dry field. It may be that the families shared sufficiently similar socioeconomic characteristics (size, wealth, and so on) that their needs and abilities to capitalize on the dry fields were relatively similar.[7] If, over time, family composition remained relatively similar, that circumstance would certainly have been conducive to villagers maintaining an equal per family distribution rule.

The rationale for conducting regular redistribution rather than simply dividing up the land permanently is a bit clearer. First, the land was located on the banks of the Yoshii River, where there was some risk of flooding, and maintenance of a redistribution mechanism allowed for sharing of any loss due to flooding. Remaining land could be remeasured, carved up into an equal number of units, and allocated to all participating families. While this appears not to have been an issue in the last decades of the system, flooding certainly had been an issue in earlier periods before modern riparian engineering. Second, as a number of discussions later indicate, getting unanimous agreement on what plots or combination of plots constitute units of equally productive land was not an easy chore. By rotating access to each plot to all families in sequence, instances where there were disagreements over the absolute yield of plots could be rendered moot. Since each family ultimately cultivated every field in sequence, the impact of any perceptible difference in field quality was negated; one simply had to wait one's turn before the opportunity would arrive to cultivate any field one felt to be marginally better.

In principle, this system could work to share the costs of flooding among the villagers in this limited section of the village. Likewise, it could be used to share the benefits of any extension of this land through poldering or other reclamation efforts. However, interviews suggest that neither happened in practice, at least during the post–World War II era.

The potential to invest in improvements in this section of the village was limited. Irrigation was not an issue for dry-field agriculture, and there was little opportunity to expand the land under cultivation through poldering into the river given its high water force. The principal opportunity to improve the land lay in the use of fertilizers and, in the twentieth century, pesticides. These investments had immediate payoffs for any farmer who used them. They did not require a long payback horizon. Any investment in machinery in either the nineteenth or the twentieth century would have done double duty; it would have been used on paddy as well as on these limited dry fields. Given the higher proportion of the farm enterprise devoted to paddy, the justification for purchase of new equipment would have come largely from its value in producing rice, not dry-field crops. Overall, there was no great disincentive to invest in the land here, but also little practical prospect.

Especially in light of the heavy investments of time that other types of redistribution could involve, it is worth pointing out that the time and labor costs of maintaining this system were minimal. Except in the case of a flood, there was no need to remeasure or reassess the productivity of the land. Further, since this practice functioned as a closed system, there was no need to repartition based on the addition of new participants. Likewise, there was no need to account for land sales, since the cultivation rights could not be traded, bought, or sold.

The low maintenance costs of this instance of *warichi* may help explain why residents continued to practice redistribution into the 1970s. However, it is also important to note again that redistribution was implemented on just a small part of cultivated land. Consequently, it represented a small part of each farm enterprise, a part designed to provide only self-sufficiency for the families cultivating, not produce to sell on the market. Every participant benefited from this subsistence safety net. At no point in my interviews did the subject of free riding arise.

The practice of redistribution ceased as a result of the national government's exercise of eminent domain over this property in the early 1970s, not because participants felt the system no longer met their needs. The Ministry of Construction determined that the Yoshii River required a new, bigger, and more modern dike to protect the surrounding region from floods. The new construction required that all of the Seiriki section of land be taken over either for the dike itself or for facilities associated with it. The land was lost to cultivation entirely.

***Per Capita Redistribution (Type II): The Case of Satsuma.*** If village-based repartition systems represent the most decentralized locus of control of land tenure, Satsuma's land tenure system, called *kadowari* (literally "gate-dividing"), marks the opposite extreme of control, one totally prescribed by the domain administration. Most domain-based systems built on village practices in the way that Kaga administrators did (see below). Satsuma distinguished itself by the early role the domain played and the extent of its dominance.

While the *kadowari* system shared an objective with Kumayama joint ownership—achieving some measure of equity in the allocation of an agricultural resource to each family—it went about it in a completely different way and with different outcomes. Although the object of redistribution was each family unit as in Kumayama, the two did not share key operational principles. Kumayama considered each family unit regardless of its component members and was not wealth-redistributing; Satsuma practice considered individual family attributes and redistributed wealth. In Kumayama, each family got land of as nearly identical productivity as villagers could manage, regardless of family size, age, and gender composition. In Satsuma, accounting for differences in family composition, notably the number of active adult males in a family, lay at the heart of Satsuma's sense of equity in allocating land to villagers. Given this structure, it did not by any means realize the socialist principle of allocating resources to each family based on their need (as discussed below), but it nonetheless redistributed wealth.

The *kadowari* system is unusual in that it consciously sought a frequent, regular periodic redistribution of wealth among families based on their size, gender, and age. In most areas that employed redistribution, it was Type III, in which allocation practices still permitted accumulation and reduction in cultivation rights based on purchase or sale, generating considerable inequality of holdings. By contrast, the *kadowari* system actively sought to redistribute wealth via a nonmarket mechanism: each family head received an allotment for every active, adult male in the household between the ages of fifteen and fifty-nine to manage.[8]

Although some commoner officials had opportunities to accumulate land to rent out, allocating arable land equally among ordinary village men limited the opportunities for tenancy to develop within that stratum of village society. Because the head of the household managed the lands of the entire family and second or younger sons of any age often remained as members of the main family, families with many adult sons could manage a farm enterprise considerably larger than a family with an equal number of members but comprising more women, retired men, or young children could manage. Such an imbalance of resources did lead to some tenancy, with families having greater needs renting lands from families with greater land management rights. Nonetheless, this imbalance was not a function of skill in managing farms and other rural business activities, nor did it result in a permanent acquisition of land rights that could be transmitted intergenerationally. The imbalance was simply a function of the luck of birth and longevity.[9]

According to historian Ono Takeo, the *kadowari* system had its origins in the 1590s. He has suggested that it represented the product of a systematic reorganization of late medieval tenurial practices. While there was no documentation, he considered the most likely first use to be 1594, the year in which Toyotomi Hideyoshi first demanded estimates of the domain's value.[10] Domainwide reallocations routinely accompanied such investigations, and there are references to *kado* in documents written within a decade of this first survey.[11] Here we have clear evidence that Hideyoshi's survey efforts failed to convey to households anything like individual, fee simple private property rights to specific plots of land. What the surveys accomplished at the village level was merely to (a) register all lands on the domain tax books and (b) assign all villages a putative yield, a value used to assess land taxes and other dues.

Under the *kadowari* system, all land in a village was divided into several categories. Most land fit into the *kadodaka* category, land subject to redistribution. Reclaimed land was always included in the *kadodaka* classification the year after it was developed and was not treated as a separate category of land. A second class of land, *ukimenchi*, was reserved for cultivation by the domain's rusticated samurai (*gōshi*), who were periodically rotated from the domain's castle town out to the villages. This was part of an effort to make some samurai self-supporting and reduce the number dependent on domain stipends.[12] Upon official recognition of some pressing need, land from this category might be exchanged with *kadodaka*. Within this second category of land, there was another (third) category, *shiakechi*, land privately reclaimed by rusticated samurai. It was not incorporated into repartitioned land unless it was exchanged for *kadodaka*. These two categories were not subject to periodic redistribution among villagers. Finally, two categories of land (*shoya ukimenchi* and *yakubunchi*) were reserved as salary lands for village headmen. These lands were separated from lands subject to reallocation and were cultivated by the labor of the lesser villagers. The special perks of rural samurai and village officials provided a means for some of them to acquire a substantial amount of land.[13]

This classification of lands removed incentives for individual ordinary farmers to reclaim land on their own. The benefits of any such effort would have been lost in the next redistribution. If land were to be reclaimed, it would be the result of a shared effort with other villagers or members of one's residential district *(kado)*, with benefits shared among all.

Each village was divided into fixed, multiple-household, geographic subunits called *kado*, which were both residential units and units of arable land reallocation and taxation. Although there were very limited exceptions, as noted below, residences were not reallocated. With the exception of village headmen's families, all types of arable were cultivated by commoners (see below). There were great variations in the number of *kado* in each village. Ono's data show some villages with as few as eight and others close to or exceeding one hundred, suggesting that the village, defined as an administrative unit, was a geographically extensive unit.[14]

The *kado* was the major unit of land surveying, tax assessment, and tax collection.[15] In these regards, it bore much more administrative responsibility than the hamlets *(aza)* that composed the subsections typical of early modern villages elsewhere. If an individual could not pay the tax due, the other members of his *kado* bore the responsibility for making up the deficit.

A village official called a *myōzu* (literally head of the *nago*, as the lesser farmers were called) oversaw each *kado*. As a result of the salary lands that came with the office, affording the *myōzu* larger holdings than other villagers, the position of *myōzu* was often bought and sold.[16] This constituted the only way to purchase or sell land under the *kado-wari* system.[17]

Participation in the reallocation of *kado* land was limited to families with active adult males. All such households were guaranteed a place in agriculture, and all bore labor duties for local as well as domain public works. While the total area of land in a *kado* subject to reallocation might remain relatively constant, demographic change in the families linked to it was continuous.[18] Of course the condition of arable land changed, too. Even if not washed away, its quality might decline, after a flood littered land with debris, for example. Just as in the other types of redistribution regimes, these changes had to be taken into account to maintain fairness in allocating tax burdens and to maintain the system's principles for allocating cultivation opportunities. Nonetheless, changes in family composition were the driving force behind reallocation.

On a domainwide scale, reevaluations of land and population took place on the occasion of each of the handful of domain surveys, but these were too infrequent to support the adjustment of cultivation rights essential to the operation of the *kadowari* system.[19] Consequently, village conditions were reviewed by *kado*, village, and domain officials at the end of each year. They inspected each village and noted the newly eligible young men, the retirees, and the condition of the land. If the land in a village changed significantly, a survey might be ordered. Based on their assessment of need, a reallocation would take place.

Allocations were made by lottery.[20] Ono suggests that each participant was entrusted with land valued at about 3.5 *koku*.[21] If one *kado* lacked an adequate supply of labor or had a surfeit, lots would be drawn to choose those men who would move to a new *kado* to relieve pressure in one and take up slack in another. In some cases, this meant relocating to a completely different village. Lands allocated at this time could not be transferred to another household unless there was a special reallocation.[22]

Except when there was a domainwide resurvey of arable land, there was no general order for redistributions to take place. There was no set interval for redistributions, although there is limited indication that if villages reallocated too frequently it created a disincentive to properly fertilize and manage fields, and the domain specifically sought to limit extremely frequent redistributions.[23] Nonetheless, the decision was left up to the *kado* in consultation with village and domain officials.

Much of the operation of *kadowari* might have been left entirely to villages, but the key exception concerned the intermittent need to transfer farmers between villages. While infrequent, it was a significant function that villages could not have accomplished on their own. That the domain paid so much attention to distributing labor over the various *kado* and villages is telling.

Domain concern with the allocation of labor reflected its interest in having all fields cultivated and therefore maximizing domain tax revenues; however, this policy also reflected other compelling thrusts in domain policy. It assured that villages had sufficient labor available to participate in both village and domain projects such as irrigation construction and maintenance, and, perhaps more important, labor for military functions. Military projects included maintenance of numerous rural fortifications and other related military duties.[24] A second consideration may have been to allocate the tax burden fairly. Inspections of harvests *(kemi)* for the purpose of calculating the land taxes due were carried out *kado* by *kado*, and the *kado* allocated the assessment among its members based on farm enterprise scale.[25] Further, regular discussions of the need for reallocations and the frequent reconsideration by officials and *kado* residents of the amount and condition of lands under the plow drew continuous attention to who was farming what lands and what the tax burdens of that land were.

Ono's study leaves some questions unanswered. Perhaps the most significant is the rationale for dividing a village up into *kado* in the first place. The presence of large numbers of *kado* in many villages indicates that there were multiple small settlements within each administrative village unit. Multiple settlements in a village tell not only of significant geographic diversity, but of considerable distance between one section of the village and another, distance great enough to make allocating land throughout the whole village to all its families simply impractical.

Redistribution procedures would have been sensitive to changes in arable due to flood, landslide, or other natural hazard, but most important were changes in the composition of families as they moved through their life cycles. Addition of land by hamlet reclamation would also have played a role in the decision to reallocate cultivation rights. Response to incidents of natural hazards was less important to the domain than distributing male labor to maximize agricultural output and taxes, and to support public works, both civil and military; thus the domain discouraged frequent reallocations to limit the disincentives to maintain fields. The inclusion of reclaimed land in the reallocations the year after its initial cultivation discouraged individual efforts and meant that the most advantageous approach to expanding arable acreage was through collaborative efforts of the *kado* or perhaps the village.

Among ordinary villagers, the only families that might potentially lose out in this system were those with few or no active adult males. In the former case, families could adopt an adult male into the family through marriage or adopt an infant boy. Those options were less likely in the case of a family that already had a male child heir, although in principle it was possible. One can also imagine that this allocation procedure placed a young bride under considerable pressure to bear male children so that over the long run the family would have access to adequate cultivation rights to support itself without renting land.

Given the heavy, omnipresent hand of domain authorities, the stimulus to employ Satsuma's *kadowari* system must be sought in the benefits that it provided to the domain. Domain policy is reflected in the system's overriding concerns, no matter how much they might overlap with ordinary villagers' interests. Reallocations took place under the watchful eyes of domain as well as village officials. Domain authorities could move cultivators to other *kado* and even villages as needed. Rights to land could not be alienated, and the opportunities to accumulate rights to land were limited to rural samurai and some of the village officials. Reclamation by ordinary farmers was largely collaborative, and while additions and losses from floods or landslides were accommodated by the system, the driving force in reallocation was changes in family structure. In Satsuma, there is no evidence the domain compromised with village traditions, even though routine implementation of the system was largely allocated to village officials, including the heads of *kado*.

***Per Share Redistribution Systems (Type III) in Kaga Domain.*** To illustrate some of the key features and consequences of the most widespread of joint ownership systems, I examine Kaga *han*, located on the western coast of central Japan (modern Ishikawa and Toyama prefectures). Its *warichi* system redistributed all kinds of village lands. Although mechanisms were developed that protected residential land of share-holding farmers, land redistribution might mean that a tenant would be forced to change his residence.[26] While such an instance was extreme and likely very rare, it serves as a reminder of the extent to which community control could be exercised over a household's usufruct.

In the sixteenth and early seventeenth centuries, Kaga *warichi* practices were almost exclusively a village matter. Ultimately, the domain administration regulated procedures for redistributing land. In both respects, the experience of Kaga domain was typical of many other domains in which per share joint ownership was practiced. It reflects the range of domain administrators' direct interest in reallocation practices from a hands-off attitude, to limited guidance, to fuller control.

The origins of the *warichi* system in Kaga domain are shadowy, but evidence indicates that the villagers themselves created the practice and arbitrated most procedures. In the early seventeenth century, domain officials were content to let villagers follow their own rules and only adjudicated disputes among villagers. By mid-century, they actively encouraged the practice as they set about reassessing the tax value of villages and implemented a variety of reforms in rural administration. However, thereafter, until the early nineteenth century, they did nothing to push the use of *warichi* or require periodic redistributions. Only from 1838 did the administrators make efforts to enforce regular use of per share proportional redistribution.

Based on work by Takazawa Yūichi, we now understand that domain villages had implemented *warichi* on their own by the very early seventeenth century, decades earlier than first thought. Other evidence confirms early-seventeenth-century village-based joint landownership, although by 1631 the domain was encouraging villages to use it in certain circumstances.[27] The first clear indication that the domain was interested in providing encouragement of land redistribution appears in one clause of an ordinance issued to all county magistrates *(kōri bugyō)* that year.[28] Beginning "in those places here and there in which land division was long ago ordered," the document indicates that in some places the domain had ordered use of *warichi*, probably in the context of village legal suits like that Takazawa uncovered concerning long-standing autonomous village practice (but not ordered domainwide). This impression is reinforced by the prime directive of this order that a land redistribution should only be implemented in accord with a consensus of shareholding villagers. Again, this evidence implies that *warichi* practices had a substantial history that domain administrative regulations reflected only partially. This ordinance also presumed the widespread existence of the practice (it refrains from specifying any particular redistributive procedural details) and the domain's long-term acceptance of it.

The 1631 ordinance also explains the relationship between land redistribution and the reallocation of retainer fiefs. Indeed, without ordering its use, this document certainly encourages it in villages that paid taxes to landed retainers rather than to the domain itself. Such cases could present special problems: in many instances, a village paid land taxes to multiple overlords simultaneously, each of whom made an independent judgment regarding appropriate taxes. Here the domain viewed *warichi* as a mechanism for allocating the independently determined tax burden of multiple retainers among all village shareholders.[29]

In principle, the practice of corporate village responsibility for land tax payments should have meant other villagers made up the difference that defaulting families failed to pay; however, it is clear that the system was not working up to expectations. The domain discerned that some farmers were paying far more than their share of land taxes—a share so large that they were unable to pay in full. In addition, retainers' agents were abusing villagers. Upset that they were not getting the tax payments they assessed, they marched into villages and resorted to physical compulsion and torture in order to try to extract payments. These difficulties ultimately led the domain to fictionalize retainer fiefs, take over land tax assessment and collection everywhere in the domain, and make erstwhile landed retainers nothing more than salaried underlings of the domain, devoid of any independent financial base.[30] In the domain's view, *warichi* provided a mechanism to maintain order, average all taxes over all shareholders, and address arrears in payments: it would result in tax assessments that assured that individual families all held the same proportion of good and bad land, of dry field and paddy, and that the entire tax burden of the village would be shared in proportion to shareholding size. This concern and the encouragement for villagers to address the problem through joint landownership marked the start of a slow, extended process by which the domain ultimately became deeply involved in regulating *warichi*.

All the preceding evidence indicates a village-based *warichi* origin that predates 1642, but from that time, domain policy makers gradually invested more of their own energies in regulating the redistributions. They first expressed interest in overseeing *warichi* practices in the mid-seventeenth century, when they specified (1651) that the district administrators *(tomura)* and the cultivation magistrates *(kaisaku bugyō)* had to give their permission before a reallocation could take place. These orders were part of the major reform of domain administration that fictionalized retainer fiefs and brought administration of them directly under control of the domain. This reform, the Kaisakuhō, also involved aggressive encouragement of *warichi* between 1651 and 1670. In this instance, authorities' concern was not just to ensure that the land tax burden was equally distributed among shareholders. It was one means by which the domain could discover lands that had been kept off the tax registers. Many of the village tax documents handed down to all villages in 1670 include a category called *teagedaka*. *Teagedaka* was the name given to new lands "voluntarily" reported to the tax office. This land was distinct from reclaimed land *(shinden)*, which had official authorization from the inception of a project. In effect, *teagedaka* was land that had been kept off the tax registers but was now being declared by the villagers before it was reported to or discovered by higher domain authorities who might penalize villages. By encouraging *warichi*, domain authorities engaged villagers in the remeasurement of the area of land under cultivation. This put farmers under pressure to declare unregistered fields without domain authorities having to send out their own surveyors.[31] Nonetheless, until the early nineteenth century, some form of natural disaster (such as flooding or landslide) or substantial change in soil fertility was the typical stimulus that generated consensus to redistribute.[32]

Yet even in the heady days of administrative reform, domain authorities declined to specify the procedures that the villagers were to follow. Like the 1642 ordinance, the domain did not mention the standards of measurement to be used, which classes of farmer were to be included in the reallotments, which lands could be exempted from redistribution, or any other procedure associated with land redistribution.[33] These matters were all left to the shareholding farmers and/or village officials to decide. The 1651 orders merely referred villagers back to earlier documents that appear to have originated in the village.

Thereafter, domain involvement in regulating *warichi* increased sporadically. In 1671, authorities specified the maximum extent to which land might be exempted from repartition, and by the nineteenth century, they licensed land surveyors.[34] In 1800, the domain first encouraged the implementation of a redistribution at least once every twenty years.[35] However, until 1838, domain authorities let villagers decide most redistribution procedures by themselves, based on mutual agreement.

In the early decades of the nineteenth century, authorities became increasingly concerned that their encouragement of redistribution too often went unheeded, and in 1838 they finally decided to take control. Villages were ordered to conduct a redistribution at least once every two decades. Villages were told how to conduct specific aspects of measurement, yield evaluation, and so forth. Other elements of measurement and yield evaluation were also discussed, including the elimination of a relatively limited category of excluded land, *sōchi*, that had crept into usage over the decades. This was especially "thin" paddy that the village claimed was owned jointly by the village. Income derived from this land was contributed to the village budget and not redistributed.[36] Even in these regulations, such important matters as the length of the measuring rod were to be determined by village custom even if that measuring rod differed in length from the official land surveying rod of the domain.[37] Just how effective these ordinances were is open to question, as some of the data below suggest, but the hands-off attitude of the domain had clearly changed irreversibly.

Especially during the early years of land redistribution in Kaga domain, there was substantial variety in the procedures employed to reallocate land. Reallocation was to be initiated solely based on a consensus of the shareholding farmers. Throughout most of the Tokugawa era, once authorities were notified and permission to proceed was received, the villagers themselves took full responsibility for determining the procedures to be followed. A number of issues were revisited at each redistribution.

Once the official permission to conduct the *warichi* was in hand, villagers had to reach agreement on the exact procedures that would be followed. These agreements *(sadamegaki)* varied in length and detail, with greater detail becoming common later in the Tokugawa era, especially after the issuance of the 1838 regulations. They specified which lands were to be redistributed and which were not. Lands subject to reallocation were called *kujichi* (literally "lot lands"); the limited portion of arable land that was exempt from redistribution was called *hikichi* (literally "pulled lands," on which more below), a designation that indicates participants in principle recognized all land belonging to the village as a corporate body.[38] In practice, *kujichi* was a residual category that included all lands not specifically exempted from redistribution.

Ultimately, the domain, too, implicitly underscored the idea that even lands that might be excluded, for example, residential garden and seed plots, belonged to the village, not to individuals who lived on it. A domain ordinance (1671) prescribed a general exemption *(hikichi)* of up to six *tan* for each hundred *koku* of assessed productive value of the village.[39] One hundred *koku* was the assessed value of a minimum of 66.5 *tan* of arable and/or residential land, so somewhat less than 10 percent of village land that could be protected from reallotment.[40] This was enough to encompass most residential compounds, seedbeds, and other similar lands that absorbed heavy investments of capital and labor.[41]

In addition to this small amount of residential and arable land, exemptions for unproductive lands in the village were also permitted. Roads, footpaths, and shrines were exempted. The small amount of land that was located in the perpetual shadow of houses, tree-lined roads, and the like, was also exempted from redistribution because it was unproductive.[42]

When the principles for defining these exemptions were settled, the village conducted its own land survey. The expenditures for the survey were borne directly by the villagers themselves. Among other costs, they had to hire a surveyor, always a commoner.[43] Also, unlike domain-run surveys, which used a very rudimentary classification of land productivity, each field was measured and its per hectare productivity was assessed in considerable detail.[44]

Once these measurements and productivity evaluations were completed, all of the land in the village was divided into sections *(wari)*, which were further divided into the subsections to which cultivation rights were assigned by lot. These field groupings were numbered and contained fields of comparable per hectare productivity. Each section was based on the natural features and characteristics of the arable land in each part of the village. The size and number of *wari* was not permanently fixed.

Three successive redistributions in Shinbo village (Noto Province) illustrate the fluid structure of the *wari* sections (Table 4.2). Total area of the village included in the 1812 *warichi* was approximately 5,545 *bu* (one *bu* is approximately four square yards). This was divided into eighteen sections. Mean section size was approximately 308 *bu*. The sections ranged in size from 45 *bu* to 540 *bu*. Shinbo farmers did not feel constrained to maintain these divisions *(wari)* in two later redistributions. They restructured them in such a way as to reduce the differences in size somewhat. The average size fell to about 298 *bu* in 1838.[45] By 1867, the area of land reallocated had increased by some 80 *bu*, but mean section size fell to 287.5 *bu*.[46] This transformation reflected substantial restructuring of the original sections. Despite the elimination of three of the sections used in 1812, there were now a total of twenty.[47] The range in size was further reduced. The smallest was now 80 *bu*, the largest only 400 *bu*.[48] With the exception of a small 125 *bu* addition of reclaimed land in 1838, all of the land cultivated in 1867 had been cultivated since 1812 (no land had been lost from cultivation). All of this reconfiguration was accomplished by rethinking the characteristics of arable land and adjusting the subdivisions accordingly.

In thinking about the possible benefits of smaller subsection size, it is important to remember that each participant drew lots for access to fields within each *wari* section of the village. Assuming no loss of land from cultivation, changes in section size per se could not affect the overall outcome of a redistribution. Any benefit to creating smaller

**Table 4.2**   Shinbo Village: Changes in *Wari* Structure, 1812–1867

|      | Wʹʀɪ Sɪᴢᴇ Rᴀɴɢᴇ | Aᴠᴇʀᴀɢᴇ Wʹʀɪ Sɪᴢᴇ | Nᴜᴍʙᴇʀ ᴏꜰ Wʹʀɪ | Cᴏᴍᴍᴇɴᴛ |
|------|------------------|---------------------|------------------|---------|
| 1812 | 45–540 *bu*      | 308 *bu*            | 18               |         |
| 1838 | 50–540 *bu*      | 298 *bu*            | 19               | 3 *wari* eliminated, 4 new ones created |
| 1867 | 80–400 *bu*      | 287.5 *bu*          | 20               |         |

*Source:* Hakui-shi Shi Hensan Iinkai 1975, *Kinsei hen*, 663.

sections *(wari)* lay in facilitating the process of dividing the village into sections of equal quality and similar characteristics. Smaller sections allowed for finer variations between sections, and it may well have facilitated reaching agreement about what lands belonged in a given section.

Before the measurement and evaluation of the land, the cultivation rights of the village had been divided into shares called *kuji* (a "lot," as in lottery). The most commonly cited hypothetical case used to illustrate this process has a village of 100 *koku* divided into ten shares valued at 10 *koku* each.[49] Those who held fractional shares were combined into a group so that their total cultivation rights equaled one full share. One of them, usually the largest holder, served as the official representative (*kujioya*, literally "lot parent") of the shares' joint owners. Holders of whole shares did not have to join with other partial holders to draw lots. In still other cases, a family might hold more than the rights equivalent to one share, and they would draw more than one lot. If, for example, they held enough rights for two and a half shares, they would draw two lots solely on their own behalf and also participate in drawing one additional lot in collaboration with other partial shareholders. (This practice was universal under per share joint ownership.)

In general, the redistribution followed these procedures: After the land was measured and its quality evaluated, each *wari* subsection was divided into the same number of subunits as there were shares *(kuji)* to be drawn. Each had the same per hectare yield. Each was identified with a number or name, and the share representatives drew lots for each subsection in every *wari* until all fields were assigned to the various holders of lots or lot-holding groups. If a lot was drawn for a group of partial shareholders, the members of that group would then hold another drawing to allocate among themselves the fields that the share leader had drawn for the group.[50] Consequently, each *wari* section would be allocated at least once and commonly twice (for multiple-member groups).

Because per share (Type III) joint ownership permitted accumulation of rights to cultivate land, tenancy was widespread. As hinted at in the preceding remark that a tenant might be forced to change residence as the result of a reallocation, tenants, too, participated. Once their landlords had determined which sections of the village they would manage, a redistribution among the tenants of each landlord took place, also by lot.

When the process was completed, each shareholder and tenant held fields of different grades and qualities in proportion to the distribution of different grades and types of land throughout the entire village. No one could accumulate only the best land. Everyone held an identically structured portfolio of rights to land. In this way, taxes could be allocated to each shareholder based strictly on the value of his cultivation rights; each would pay the same tax per unit of assessed value of land. There was no need to vary tax rates based on land quality, type, or area.

Again, nothing in this per share redistribution system inhibited the accumulation of rights to cultivate many hectares or constrained the loss of rights to land. Shares were subject to purchase, sale, inheritance, and transfer in the case of failure to repay a loan for which cultivation rights were pledged as security. In Kaga, as in other regions practicing Type III ownership, great inequality in land rights was common. When shareholders participated in a redistribution, wealth was not redistributed. Participants got out of the redistribution rights to the same percentage of village land as they had prior to it. Proportionally, they lost nothing. If the actual area they cultivated changed, it was solely because the total arable land under cultivation in the village had been increased by land reclamation or decreased as the result of flood, landslide or other disaster. Initial inequalities in shareholding within the village were replicated in the outcome.

Incentives were maintained to invest in land reclamation. In many instances land reclamation was undertaken as a joint village project, and access to its fruits were shared in proportion to individual investments in the project. Sometimes the reclaimed land constituted a *wari* section in the village on its own, but even if it was combined with other land, the size of a participant's share was increased in proportion to his investment in the project.

If land was reclaimed by an individual, provision was also made to secure for him the results of his labor. Such land was often excluded from a redistribution for several years. Even when the reclaimed land was folded into the regular redistribution, his share of the village land was increased accordingly, and while the next lottery might remove him from cultivating the land he had reclaimed, he would draw lots to cultivate the amount of land he had added to the village. So if a village consisted of ten shares giving access to 100 *koku* of land and one individual holding one share reclaimed an additional *koku*, when the reclaimed land was redistributed, the ambitious farmer would draw for 1.1 shares, not just his original single share.

During the early seventeenth century, the village-based nature of land redistribution encouraged diversity in the incentives embedded in *warichi* practice. In some cases, such as those from Noto's Nafune district, original fields *(honden)* were divided quite equally among shareholding farmers. Differences in the shares held by each household appeared solely as a result of their initiative in reclaiming arable land.[51] In other villages, the scheme found in Ota village during the redistribution of 1657 may have been used. Here, the best quality land was distributed among shareholders in proportion to the total value of the holdings of each. Average and low-grade paddies were distributed among the shareholders at flat rates of 100 *bu* and 60 *bu* each.[52] In still other cases, good, average, and poor quality land all were allocated to each shareholder in direct proportion to the presence of each grade of land in the village, using practices similar to those described above. This format ultimately came to dominate *warichi* procedures during the mid-seventeenth century and after.

To this point I have outlined the general principles for Kaga *warichi* and the contours of its development from a village practice to heavy domain involvement. I next look at intervals between redistributions as a means of understanding the frequency of their use and as a basis for some initial thoughts on the stimuli for villagers to implement a redistribution at their own volition, not on domain initiative. Again, there are no domainwide data on this subject, so we must rely on scattered examples of actual practice.

Domain ordinances cited above clearly indicate that *warichi* redistributions be conducted when there was a consensus in the village that one was needed. In the years before the Kaisakuhō reforms of the mid-seventeenth century, there were two general stimuli that might create such a consensus: first, administrative problems created by the differential tax assessment of multiple retainers, or encouragement and mandate by the domain; and, second, changes in the area or productive value of village farmlands.

After the mid-seventeenth century, the problems created by multiple retainer taxation of a village were removed as an issue. While the domain actively encouraged *warichi* in its mid-seventeenth century reforms, it was not required. Thereafter, down to the late eighteenth century, the domain made no effort to encourage or mandate the practice. So these administrative stimuli to create a consensus played a relatively small role over the Tokugawa era as a whole.

What of the second category of stimulus, changes in the area and quality of land cultivation? Changes in the quantity and quality of arable stemmed from two different sources. On the one hand, villagers individually or jointly, as part of a domain-planned project or in collaboration with other villages in a district, invested in extending the area of land under the plow or increasing its productivity through improved irrigation and/or drainage. By the late seventeenth century, almost all of the opportunities for large-scale reclamation of agricultural land had been exercised.

While small-scale reclamation projects continued throughout the Tokugawa era, loss of land from productivity, efforts at restoring land, and declines in productivity were far more significant than new reclamation. Many of the mid-Tokugawa reclamation projects simply restored recently damaged lands to productive use rather than opening virgin lands. Thus, for the period as a whole and for the middle seventeenth century and later half of the era in particular, the primary stimulus for a redistribution lay in the loss of land or significant changes in its productivity.

How sensitive was Kaga *warichi* practice in responding to loss of land or productivity? A look at long-term intervals between redistributions in a village provides some insight. They are not a perfect measure since we do not have annual documentation of floods or landslides for these villages. But in combination with domain policies enacted in the nineteenth centuries, we can paint a reasonable picture of how sensitive Kaga village practice was to natural hazard risks.

We have data for eight villages (Table 4.3): Kitajima, Chōkeiji, Hongo, Shinbo, Tsubouchi, Myōga, Takasu-de, and Jūne'naki. The data may not represent a complete record of redistributions for each village. Nonetheless, they have the virtue of presenting villages from very different parts of the domain, from Toide in what is now Toyama Prefecture, to villages on the Noto Penninsula near Kanazawa, to just south of modern Kanazawa City.

Based on these cases we can conclude that the actual interval between redistributions varied substantially. The villages in Table 4.3 show intervals as short as eight years or as long as eighty-three years. The average interval for these villages varied from one occurrence every fourteen years to one every forty-two years. The longer average was clearly the product of unusual circumstances: only one village experienced an average interval that long. Again, bear in mind that the domain did not formally require *warichi* to be implemented at least every twenty years until 1838. Even before that time, however, the most frequent intervals fall between fifteen and twenty-nine years. In this

**Table 4.3**  Intervals between Reallocations for Eight Villages

| | Kitajima | | Chōkeiji | | Hongo | | Shinbo | |
|---|---|---|---|---|---|---|---|---|
| | Redistribution Year | Redistribution Interval (years) | Redistribution Year | Redistribution Interval (years) | Redistribution Year | Redistribution Interval (years) | Redistribution Year | Redistribution Interval (years) |
| | 1672 | ? | 1785 | ? | 1666 | ? | 1722 | ? |
| | 1694 | 22 | 1801 | 15 | 1680 | 14 | 1749 | 27 |
| | 1730 | 36 | 1816 | 17 | 1691 | 11 | 1768 | 19 |
| | 1762 | 32 | 1838 | 20 | 1718 | 27 | 1787 | 19 |
| | 1805 | 43 | 1856 | 18 | 1736 | 18 | 1816 | 29 |
| | 1826 | 21 | 1871 | 15 | 1758 | 22 | 1838 | 22 |
| | 1841 | 15 | | | 1795 | 37 | 1867 | 29 |
| | 1853 | 12 | | | 1814 | 19 | | |
| | 1865 | 12 | | | 1838 | 24 | | |
| | | | | | 1857 | 19 | | |
| | Average | 24 | Average | 27 | Average | 21 | Average | 24 |

| Tsubouchi | | Myōga | | Takasu-de | | Jūne'naki | |
|---|---|---|---|---|---|---|---|
| Redistribution Year | Redistribution Interval (years) | Redistribution Year | Redistribution Interval (years) | Redistribution Year | Redistribution Interval (years) | Redistribution Year | Redistribution Interval (years) |
| 1733 | ? | 1729 | ? | 1772 | ? | 1734 | ? |
| 1753 | 20 | 1769 | 40 | 1786 | 14 | 1817 | 83 |
| 1776 | 23 | 1791 | 22 | 1794 | 8 | 1839 | 22 |
| 1797 | 21 | 1811 | 20 | 1810 | 16 | 1859 | 20 |
| 1817 | 20 | 1831 | 20 | 1829 | 19 | | |
| 1837 | 20 | 1851 | 20 | 1841 | 12 | | |
| 1857 | 20 | 1871 | 29 | 1853 | 12 | | |
| Average | 21 | Average | 24 | Average | 14 | Average | 42 |

*Sources*: Kitajima and Chōkeiji: Tochinai 1936, 106–108. Hongo and Shinbo: Hakui-shi Shi Hensan Iinkai 1975, *Kinsei*, 663. Tsubouchi, Myōga, Takasu-de, and Jūne'naki: Toide Chōshi Hensan Iinkai 1972, 371.

regard, the domain's choice of a mandatory maximum interval appears to be moderately close to intervals seen in these cases. After 1838, this sample indicates that domain policy of redistributing lands about once every twenty years was followed generally but not universally.

The variation in the length of intervals between reallocations suggests that the *warichi* system showed some responsiveness to the impact of natural disasters. Again, since we do not have data on floods and landslides, or data on changes in the status of irrigation facilities or other factors that affected soil productivity, this conclusion must be tentative. However, the common intervals reflected in these data are short enough to indicate the possibility of moderate responsiveness.

While *warichi* promoted the common good by creating a foundation for fair taxation, it required considerable effort on the part of villagers, and a bigger issue remains: whether the longer intervals reflect a lack of need due to stability in the area and quality of land, or whether they reflect an inability of shareholders to reach consensus as to the need, a problem that could arise from the opposition of just a small number of villagers when changes were more subtle than those induced by a significant natural hazard. The variation evident in this small sample does hint that there could difficulty in getting consensus within a village to implement a redistribution. Even if people recognized a need for some adjustment, they might well argue for an alternative solution. Some villagers might have argued that, rather than conduct a redistribution, simple adjustments could be made to the way in which land taxes and other dues were allocated in the village. In other words, they might have preferred a low-labor, desktop solution. Their reluctance to redistribute would not have been entirely unreasonable. Having farmed the same fields for a number of years, shareholders would also have come to feel they understood the peculiarities of their individual fields, their *kuse*. Conducting a redistribution would have meant learning the peculiarities of new sets of fields, something to be avoided unless a clear, larger objective were served by redistributing lands. Thus many villages took precisely this accounting shortcut in the face of relatively small changes in land area and quality. Under circumstances like these, creation of a shareholder consensus to reallocate land was likely an arduous and potentially contentious process.

Assuming that our records are reasonably complete for each village, the longer intervals likely reflect the challenges of achieving a shareholder consensus to redistribute. The longer intervals in these tables suggest inconsistent standards for implementation. Villagers were apparently not conducting redistributions on their own as frequently as may have been warranted by conditions, or not as often as domain policy makers thought they should. Although the domain authorities felt moderately frequent redistributions desirable, in the face of gradual, relatively subtle changes over time, village decision-making mechanisms did not live up to policy maker expectations. The domain's encouragement of a twenty-year interval in 1800 substantiates this interpretation of these data. It also suggests why they ultimately enacted a twenty-year maximum interval between redistributions in 1838.

For most of the Tokugawa era, Kaga *warichi* practices made free riding virtually impossible. For all but the last decades of the Tokugawa era, reevaluation of land under the plow and conduct of a redistribution was not periodic but depended on consensus of villagers, and thus no farmer could predict when the next redistribution would occur. Absent any ability to predict the onset of a reallocation, only an inveterate gambler or slouch would choose to reduce inputs of fertilizer and labor in anticipation of a redistribution.

To the extent that the system was flexible, it better permitted the shareholders and tenants to share the impact of changes in soil fertility and natural hazards. Their exposure to the economic hardships imposed by nature was significantly limited by this practice. In this sense, *warichi* helped shareholders to deal with both economic marginality and variation in average tax rates per hectare that developed as soil quality improved or fell with the passage of time. Yet to more accurately determine the responsiveness of Kaga's per share proportional redistribution to natural hazards would require a reasonably complete record of such events, the extent of damage they caused, and complete redistribution records for a sample of villages—materials not currently available. (I will revisit this issue using data from Echigo and a somewhat different methodological approach.)

In sum, domain interest in promoting *warichi* was threefold. First, it encouraged villagers to police themselves and to encourage full registration of all arable on the tax rolls. Hence the domain encouraged, and later required, that the system be used. Second, *warichi* limited the possibility that some farmers would go belly up and leave the villages as the result of changes in arable land area or quality. Microclimatic and natural hazard risk were shared equally by all in the village. Finally, in principle it served as a fair but simple basis for allocating land taxes and other obligations within a village. Since all villagers held the same diversified portfolio of cultivation rights, taxes could be simply assessed based on the value *(kokudaka)* of one's holdings.

An issue implicitly raised here concerns the motivation of large landholders. Why would they consent to this practice? What did they stand to gain from supporting it? One benefit may have been found in the ability of the system to simplify allocation of taxes, a responsibility that generally fell on the more well-off villagers who served as village headmen and council members. A more important concern may lie in the use of the reallocation process by their tenants. Just as the domain saw benefits in assuring that farmers survived to pay taxes fully, landlords had an interest in minimizing the problems tenants might have in paying rents. While not a guarantee, structuring tenants' cultivation access in the form of a diversified portfolio of lands through redistribution made a significant contribution to the stability of the tenant farming enterprise and, with it, the landlord's revenue stream.

## The Extent of Redistributional Practices

One of the major conundrums regarding joint ownership practices lies in assessing the extent of joint landownership in Japan during the early modern era. Although it is widely thought to be a rare practice, evidence suggests that joint ownership was used to control land rights in a significant, albeit minority, proportion of Japan's farmland during the early modern era. Estimating the location and extent of *warichi* is a challenging task because there was no national decision-making authority that developed policies on land tenure systems. The lack of national administrative control also meant that no one was in a position to compile nationwide statistics on the use of joint forms of landholding.

Nonetheless, given the scholarly emphasis on broadly uniform patterns of individual ownership/use rights over specific plots of land, it is important to make some assessment of the extent of joint ownership practices.[53] The scholarship on late-sixteenth-century land surveys provides some sense of the reach of these practices, but there has been no effort to sketch the extent of specific tenurial regimes. Such estimates are inherently quantitative, and, while any effort is imperfect, it is important to make some explicit effort to gauge their extent. This is especially so given that the presumption of a link between land survey register format and land ownership type—the listing of individual names by individually described plots that scholars widely understand as evidence of individual ownership rights to land—does not always hold. That is, register formats that are taken as evidence for fee simple ownership are also found in villages that practice joint ownership.[54] Two approaches yield instructive, although static, estimates of the extent of joint holding of arable land.

Based on prewar studies alone, there was a substantial body of evidence to indicate that the practice of land redistribution was a widespread phenomenon. Furushima Toshio's review of the literature up to the early 1940s listed studies of the *warichi* system in twenty-two of Japan's sixty-six traditional provinces as well as Okinawa, the Ryukyu Islands, and Taiwan.[55] Aono Shunsui's more recent studies added at least one province to this list, Iwaki.[56] Within these provinces, domain governments either established or permitted villages to establish land reallocation systems. Because the initiative for these systems was apparently local and sprang from local problems, there was substantial variety in the date of origin and the length of time over which redistribution was practiced. Figure 4.1 shows that while these regions may have been "exceptions" to general land tenure practices, the systems practiced in these areas were certainly widespread and not minor. Furthermore, while a number of these provinces are located in the backwaters of Japan, there is no simple way to write this phenomenon off as a product of economic backwardness. Provinces from the more densely populated, economically diverse, and commercially oriented regions of Japan are well represented in this list.[57] Note that I have omitted the Ryukyus including Okinawa from this map although scholars believe that *warichi* was widely practiced there. (Whether the Ryukyus should be considered a part of Japan at this time is open to discussion. The island chain occupied an ambiguous position under the suzerainty of Satsuma domain. However, China also claimed the Ryukyus, and the Ryukyu Kingdom received emissaries from China as though it were a part of Chinese territory.) Hokkaido is also omitted since it was not incorporated into the Japanese realm until the nineteenth century.

The chief value of Figure 4.1 is that it helps give a sense of the geographic distribution of systems of joint control over arable land. The practice appears in widely scattered and very diverse regions throughout Japan. The map also demonstrates that, based on present knowledge about the distribution of *warichi*, it was notably absent from the San'yo and San'in districts southwest of Kobe and from the area of modern Tokyo and its hinterlands.

One must be cautious in using this map as an estimate of the proportion of early modern Japan that practiced joint tenures on arable.

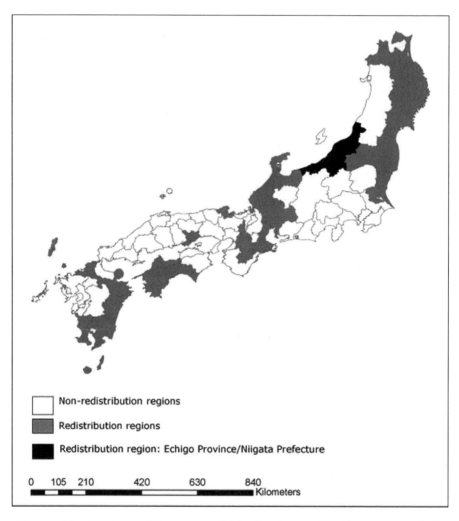

**Figure 4.1** Provinces of Documented Joint Landownership. Provinces in which land redistribution practices can be documented are shown in gray and black. The twenty-six provinces are Bizen, Bungo, Chikuzen, Echigo, Echizen, Etchū, Mutsu, Hitachi, Hyūga, Iki, Ise, Iwaki, Iyo, Kaga, Kawachi, Mino, Noto, Osumi, Owari, Rikuzen, Satsuma, Tango, Tosa, Tsushima, Yamashiro, and Yamato. Not shown is Okinawa. To this map one might add the Ryukyu Islands stretching south of this map toward Taiwan. The Ryukyus saw widespread use of joint corporate land tenure over arable land. (As noted in the text, provinces were not administrative units. Only some villages in each may have used *warichi*.) (Sources: Furushima 1939; Aono 1982; Kumayama-chō Shi Hensan Iinkai 1993)

From this perspective, it is rather inaccurate: if provinces had been actual administrative units during the early modern era, one might conclude meaningfully that redistribution was practiced in more than a third of Japan's premodern provinces; however, they were not government divisions. Tokugawa administrative divisions typically did not correlate at all with the traditional provincial boundaries. Daimyo domains, shogunal lands, and the lands of lesser shogunal retainers *(hatamoto)* frequently comprised less than a province or crossed provincial boundaries. Consequently, we need some additional device to assess the extent of *warichi* in early modern Japan.

As an alternative estimate, I calculate the official putative yield values *(kokudaka)* of those daimyo domains that mandated and regulated the implementation of joint control of arable at some time during the early modern era. In each of these provinces, domain laws regulated the use of some form of joint control and redistribution. As the case of Kaga showed, domain mandates and regulations at one time did not preclude village-based initiatives at other times. The upper sections of Table 4.4 show domains affected at some time during the early modern era, but the table does not indicate the duration of the practice.

Secondary literature does not provide data on duration for some of the cases noted here, but the data we do have suggest only moderate fluctuation. Of those for which we have information on duration, the case of Fukui's domain-controlled joint ownership is highly exceptional in that domain-sponsored *warichi* was attempted only briefly (individual villages employed it their discretion at other times). Kaga, Satsuma, Tosa, and Hizen are more typical in that they encouraged or enforced joint ownership practices throughout most of the Tokugawa era. While there surely was change over time, it appears generally to have had a modest impact on the nationwide picture.

These data, too, are not without problems. There was no consistent standard for enumerating putative yield during the early modern period.[58] Early nationwide data are estimates, and as a base figure for nationwide assessed value I employ 18,000,000 *koku*, a widely accepted figure for Japan's *kokudaka* ca. 1600. In addition, as the 1690 estimate for Tsushima (column 4) indicates, domain putative yield values were not always calculated just on the basis of agricultural output. While Tsushima was extreme, it was not unique.

Furthermore, as a glance at the table makes clear, these sources are not entirely consistent. Domain *kokudaka* are drawn from three different sources: two versions of a widely respected historical dictionary

**Table 4.4** Assessed Value of Seventeenth-Century Domains Regulating/Mandating *Warichi* According to Multiple Sources

| DOMAIN | PUTATIVE YIELD (*KOKUDAKA*) | | | | REMARKS |
|---|---|---|---|---|---|
| | (1) | (2) | (3) | (4) | |
| Fukui | 670,000 | 670,000 | 250,000 | 300,000 | |
| Fukuoka | 523,100 | 523,100 | 523,000+ | 600,000+ | |
| Fukuyama | 100,000 | 100,000 | 100,000 | 147,000 | |
| Hiroshima | 498,000 | 498,000 | 426,000 | 580,000 | |
| Hizen (Saga) | 357,000 | 357,000 | 357,000 | 500,000 | |
| Iki (part of Hirado domain; domain-mandated *warichi* limited to Iki) | | | | | 36,800 *koku*; not listed in any of the four sources; source: Iki city government |
| Imabari | 220,000 | 220,000 | 40,000 | 60,000 | |
| Kaga | 1,195,000 | 1,195,000 | 1,200,050 | 1,600,000 | |
| Matsuyama | 200,000 | 200,000 | 150,000 | 203,000 | |
| Moriyama | | | | | 20,000 *koku*; created Genroku 13/1700 |
| Owari | 539,500 | 473,344 | 619,500 | 1,000,000 | |
| Satsuma | 612,500 | 609,500 | 729,000 | 1,000,000 | |
| Tosa | 202,600 | 202,600 | 222,600 | 232,600 | |

| | | | | |
|---|---|---|---|---|
| Tsushima | 10,000 | 10,000 | 20,000 | 100,000 | four-fifths of the 100,000-*koku* figure includes the value of trade revenues to *kokudaka* |
| Uwajima | 81,000 | 80,000 | 100,000 | 100,000 | |
| Total | 5,208,700 | 5,138,544 | 4,267,150 | 6,422,600 | |
| National *kokudaka* total | 18,000,000 | 18,000,000 | 18,000,000 | 26,430,000 | Genroku *Gochō*, compiled 1700–1703, as cited in Kikuchi Toshio 1977, 137 |
| % national *kokudaka* | 28.94 | 28.55 | 23.71 | 24.30 | |
| Echigo Province | 600,000 | 600,000 | 600,000 | 600,000 | |
| Total with Echigo | 5,808,700 | 5,738,544 | 4,867,150 | 7,022,600 | |
| % national *kokudaka* | 32.27 | 31.88 | 27.04 | 26.57 | |

*Note:* Column headings indicate the following sources: (1) Takeuchi Mitsutoshi and Takeuchi Rizō, *Kadokawa Nihon-shi jiten*, 1165–1193 (early seventeenth century); (2) Asao Naohiro et al., *Kadokawa Shinpen Nihonshi jiten*, 1293–1327 (early seventeenth century); (3) Kanai Madoka, *Dokai kōshū ki* (base *kokudaka*, early to mid-seventeenth century); (4) Kanai Madoka, *Dokai kōshū ki* (Genroku 3/1690).

*Additional sources:* Aono 1982; Furushima 1939.

for early-seventeenth-century data and two sets of data compiled in a late-seventeenth-century source, the *Dokai kōshū ki* (1690).[59] The latter includes "base" *kokudaka* for each domain that appears to represent its putative yield as of approximately 1640, and estimates of domain size as of about 1690. My purpose in presenting different estimates is not to look at a trend over time, but to use multiple independent estimates of the size of domains that regulated/mandated *warichi* in order to calculate their relative weight as a proportion of the assessed value of Japan as a whole.[60]

Domain locations and assessed values often changed during the two and a half centuries of the era, but, since joint land tenures were generally implemented early in the seventeenth century, I have selected assessed values from that period to populate the first two columns in Table 4.4. Given that a number of these domains increased in size in the late seventeenth and eighteenth centuries, the table presents generally conservative estimates of the assessed value of domain-regulated joint landholding practices.

Despite problems with the data, a consistent range of 24 to 29 percent of the official value of Japan's domains mandated and regulated joint village management of arable land.

Readers should bear in mind two caveats: Again, each domain listed may only have mandated use of joint ownership for part of the early modern era, sometimes for only a short period of time, as noted above for Fukui domain, which standardized and mandated widespread village practice for only a single year, Kansei 12 (1800), then reverted to simple village-level control.[61] Other domains, such as Satsuma, mandated joint ownership for the entire period, and they are more typical. Second, the existence of a structure for regulation and laws compelling use of joint ownership say nothing about the level of enforcement. I have discovered copies of laws for only one domain, Tosa, for which there is no Tokugawa evidence regarding enforcement. That said, while traveling through Japan on my preparatory survey, I spent time in Kochi, assessing the possibility of making it a case study. Local archivists and historians indicated that there were no known village-level documents that dealt with actual implementation of joint ownership, and some even questioned if there was any enforcement at all. However, one of my consultants at the Kochi Prefectural Library recalled that during World War II residents in his community had drawn on older residents' experience and resuscitated a version of the domain system—suggesting that it was at least employed in some locations.

Tosa aside, in the other examples cited in Table 4.4, evidence testifies that laws were enforced with some degree of rigor.[62] Further, given the scholarly emphasis on Hideyoshi's edicts regarding land surveys and their nationwide effect, even the existence of codes like Tosa's indicates that daimyo did not mechanistically accept any implication that survey registers conveyed private rights to village cultivators.

As discussion of redistribution patterns above has noted, domain-based forms of joint landholding were not the only widespread pattern, and, consequently, the estimates in the upper part of Table 4.4 have one additional shortcoming: they exclude those areas in which villagers, completely independent of domain policy, implemented a joint landholding regime. This was true, for example, in a number of villages located in Tokugawa house lands, domains like Fukui, Tōdō, and more. These communities created their own regulations even though their overlords never made any effort to promote, restrict, or regulate this practice. Since the structure of the Tokugawa state meant that there was no one to keep track of nationwide use of landownership practices, we have no systematic data on the extent of village-level *warichi*.

Consequently, I had no choice but to try to create a proxy value. The proxy I employ is the assessed value *(kokudaka)* for Echigo Province listed at the end of Table 4.4. This is, I believe, a conservative estimate of villages practicing *warichi* entirely on their own. Echigo's total assessed value was 600,000 *koku*, a bit more than 3 percent of the early-seventeenth-century 18,000,000 *koku* national assessed value employed by the first three sources for the table.

Echigo's *kokudaka* is an appropriate proxy for all village-based *warichi* for several reasons. First, Echigo itself witnessed high levels of village-based redistribution, likely up to 70 percent of the province's villages, representing approximately 420,000 *koku* of the province's 600,000 total. Since Echigo was under several different administrative jurisdictions and no domain-level were records kept regarding the extent of joint landholding, this estimate is conjectural; however, it is an informed conjecture based on extensive conversations with local historians who have worked for decades studying the region's extensive central and southern (Chūetsu and Jōetsu) village manuscript collections. In all but the northernmost two counties, joint landholding was widespread. In the two northern counties (Iwafune and Kita Kanbara), scholars have yet to find evidence of joint holding of arable land.

Assuming an average village assessed value of 500 *koku*, the remaining 180,000 *koku* (30 percent of the province *kokudaka*) in my proxy estimate would represent the assessed value of about 360 villages at a time when Japan comprised some 63,500 villages, a very small fraction. In addition to widespread use in Fukui, other published case studies of village-based joint landholding reveal numerous such villages outside Echigo. In addition, based on my own fieldwork, I believe there to be considerably more such cases. Without attempting a massive, ground-level survey, I have turned up instances for which I can find no published discussion and in regions not known for using *warichi*—in modern Okayama and Saitama prefectures, for example. Other cases are likely to be discovered, so treating the entire Echigo *kokudaka* as a proxy for total value of land under village-based *warichi* is still a conservative initial estimate.

Using the 600,000 *koku* estimate for village-based joint land tenures and adding it to the figure for domain-based *warichi*, the total assessed value of arable under the sway of joint regimes is consistently in the range of 27 to 32 percent of the assessed value of the country. (Again, for the reasons explained above, Hokkaido and the Ryukyus have been excluded.) Whichever estimate and data source is employed, it is clear that joint control of arable land was a minority practice that was nevertheless extensive. Despite caveats about this estimate and problems with the data sources, this conclusion would remain valid even if the estimate were cut by half or more. My purpose here is not to insist on the accuracy or precision of a specific estimate, but simply to make the point that joint ownership comprised a significant number of Japanese villages at some point during the Tokugawa era, upwards of a quarter to a third.

## Conclusions

The preceding discussion has illustrated the major types of joint ownership of arable land: per family, per capita, and per share allocation. It has described the array of procedures and effects that can be seen in the historical record as well as the levels of control that shaped land tenure systems.

In Seiriki, the key example in Kumayama, villagers entirely on their own developed a per family redistributive mechanism based on sequential rotation. An unusual set of circumstances—the dearth of dry-field lands—sparked joint control and meant that only dry fields were reallocated. This land could not be bought or sold, and while adjustments in land could be accommodated for reclamation or flood loss, that was a secondary effect; fairly allocating access to a scarce resource was primary. Tax fairness, free rider issues, and incentives to invest in expansion were relatively unimportant.

Satsuma exemplified a purely domain-inspired system, a per capita allocation of all arable lands based on family composition. Villagers had no apparent input into most aspects of the system. They could determine the timing of reallocations, although that decision was routinely shared with domain representatives and reallocation was compulsory when domainwide land surveys were undertaken. While the *kadowari* system may have benefited villagers, the domain's primary motivation was to maintain a locally available supply of labor for public works and military support. The domain's objectives were paramount, not villagers'. There is no indication that rotation as practiced caused significant free-rider problems. Natural hazard risk could be accommodated, but the system responded primarily to changes in family composition. Incentives were present for joint cooperative reclamation projects by a *kado* but not for individual efforts. Except for village headmen, accumulation of land rights was not possible, although some degree of tenancy did exist.

Kaga's per share *warichi* lies in between the Satsuma and Seiriki practices. *Warichi* originated in villages, where participants were concerned about fair allocation of taxes in unstable geographic and climatic circumstances, even if changes were gradual. All arable lands in a village were redistributed. Unlike Seiriki or Satsuma, Kaga shareholders could accumulate or lose shares through a combination of inheritance and market mechanisms. Landlordism coexisted with redistribution. In the middle of the seventeenth century, the domain began to encourage redistribution to promote increased domain revenues and full land tax payment. Only in the late eighteenth and early nineteenth centuries did the domain increasingly regulate aspects of village practice, licensing surveyors in particular. In 1838, it ultimately required a redistribution at least once every two decades, hoping to assure full payment of land taxes and formal registry of reclaimed land. Domain initiative and data on intervals between iterations indicates that, at least in cases of modest changes in yields or area cultivated, villages had difficulty achieving a consensus to conduct a redistribution. Intervals between redistributions were unpredictable until 1838, so it would have been risky for a family to attempt to free ride. There was no knowing when the next redistribution would occur, no ability to predict when one could free ride and let the next cultivator bear the consequences (presuming one did not draw a lot for a section one had cultivated up to the time of the redistribution!). Incentives were maintained for both individual and joint land reclamation projects.

Kaga data on redistribution intervals as well as the history of domain policy suggest it could be very difficult to reach agreement to reallocate and indicate the necessity for some individual or group in a village with sufficient persuasive or coercive powers to manufacture a consensus. Absent domain-mandated reallocations, there could be extended periods without a reallocation of cultivation rights, suggesting the difficulty in getting a consensus to redistribute. Domain frustration with long intervals reinforced this impression. Calls to redistribute surely relied on practical appeals that reallocation facilitated each cultivator's full payment of land taxes or contribution to villagewide public works but may also have involved appeals to villagers' sense of mutual responsibility. Such appeals alone were not necessarily sufficient to carry the day since there were alternative ways to handle modest changes in the area and quality of arable land. Someone needed to craft a consensus in support of one alternative or another.

Incentives to improve land were structured into each of these forms of joint landownership, although not in equal degree. There is variation across types but also variation within type. In this regard, the different mechanism inducing a reallocation of cultivation rights was critical in per share systems. One must look at two different aspects of agricultural activity in relation to incentives to invest in improvements. On the one hand are incentives that are applicable to individual effort; on the other are those that require cooperation among farm enterprises. In the context of Japanese agriculture, the former include activities such as crop rotation to restore nutrients or maximize potential for double-cropping, applications of fertilizer (night soil, green manure, or commercial), maintenance of paddy beds, ridges and irrigation gates opening on one's fields, weeding, and even individual efforts to expand the area of land under the plow. Activities that require cooperation with others include collaborative expansion of arable land and development, and maintenance and operation of water control infrastructure, most directly, shared irrigation systems. Incentives for any system of ownership may operate differently for each of these aspects of farming

In each of the joint ownership regimes, cooperative village improvements—improved irrigation, drainage, land reclamation—were certainly not discouraged and represented a common approach to land improvement throughout the Tokugawa era. While it is possible that individual incentives to improve land were limited under Seiriki's per family redistribution regime, that limitation would have affected investments in expanding arable, not increased investments in fertilizer, weeding, and other investments that paid immediate dividends. Satsuma's *kadowari* system created disincentives for individual farmers to expand arable land they managed; however, incentives for village investments to reclaim land were present. It is much harder to infer individual disincentives for the per share redistribution pattern found in Kaga domain. For most of the Tokugawa era, the trigger for redistribution was not a regular interval as in Seiriki or the annual reevaluations evident in Satsuma, but a consensus among shareholders that a reassessment and redistribution of village arable was necessary. No one could know when the next redistribution would occur, and evidence introduced above suggests the interval could be quite long. Further, individual efforts at reclamation resulted in an increase in a family's cultivation rights. In Kaga, in addition to any benefits from investments to bolster soil fertility or reduce damage from insects, investments in expanding arable clearly reaped rewards. Throughout Japan more broadly, not all per share redistributions were triggered by shareholder consensus, nor did they necessarily take place infrequently. Some, as the analysis of Echigo *warichi* below shows, occurred at predictable intervals that could be as short as one or two years. Perhaps incentives to invest in expansion of arable under such arrangements functioned differently. However, Kaga's case demonstrates that there are circumstances under which joint per share ownership rewarded individual family investments in land, and to that extent it may be compatible with increased cultivator involvement with the market.

This review of different forms of joint ownership also opens a crack in the purported link between sharing natural hazard risks and the presence or absence of this form of landholding. While per share systems have the potential for such a link and some (e.g., Kaga) explicitly used a flood or a landslide as a trigger to redistribute, that was not true for all types of joint ownership analyzed here. In the per family redistribution regimes this was a secondary stimulus, with routine redistributions achieving some sort of equity among families as it bore on their ability to produce vegetables. Per capita redistribution in Satsuma may also have had outcomes that ameliorated the impact of floods and landslides, but that was not its main focus. Its chief outcome was to distribute labor to maximize use of arable lands and retain village labor for domain purposes.

While a minority practice, joint landownership was present in a significant part of early modern Japan. The preceding estimates of *warichi*'s national distribution and extent as a proportion of arable represent a reasonable, useful starting point for understanding *warichi*'s place in Tokugawa Japan. They indicate that joint ownership appeared widely, in diverse economic and geographic circumstances.

Turning from the general to a local case study offers the opportunity to explore more closely issues such as the relationship between natural hazard risk, soil variability, and the operation of per share joint ownership systems. Echigo is a region that differs from Satsuma and Kaga by virtue of a complete lack of domain involvement in joint ownership customs. This characteristic allows us to look at village decision making in redistribution without the need to consider domain motivations. The Echigo case helps us to understand the level of village commitment to this practice and fairness in its operation, and allows us to explore its adaptability to local conditions, including commercialization of agriculture.

# 5

## Lay of the Land:
## *Warichi* Practice in Iwade Village

Previous chapters have examined the general structure of *warichi* systems and the broader historical contexts within which they operated. They have also summarized the scholarly evaluation of *warichi*'s origins. It is now time to look closely at how the system functioned in practice at the village level. Did participants actually measure and reevaluate fields routinely? Did the distribution process adhere to principles of fair play in practice? Did the system adjust for changing productivity over time?

In Chapter 3 a variety of evidentiary issues were noted that complicate analysis of *warichi* systems and their operation, particularly over extended spans of time. The Satō Family Documents at the core of the analysis in this chapter and Chapter 7 provide a rare collection of relevant village documents that overcome some of these limitations. In particular, they describe repeated iterations that provide a sense of development and transformation over an extended period of time. Describing *warichi* in Iwade Village during the eighteenth and nineteenth centuries, these documents record periodic redistributions over a period of 138 years (1710–1848). While reasoning from the outcomes revealed in these manuscripts still has its limitations, the long-term perspective they provide is unique.

As in many examples of land redistribution, the precise origins of the practice in Echigo are not clear. According to two Tokugawa era gazetteers, *Echigo fudo kō* (A cultural geography of Echigo) and *Onko no shiori* (A guide to investigations of the past), the first redistributions were associated with 1610 land surveys of the Shibata domain. *Warichi* apparently developed as part of village efforts to reduce intravillage quarrels that resulted from villagers' perceived inequities in the surveys. These inequities grew out of the differences between assessed land values implemented at that time and actual productivity of the land as villagers understood it.[1] Shibata represented a fraction of Echigo, and origins here do not explain presence of the practice in other parts of the province, but by whatever process, joint landownership practices became a widespread, permanent feature of many Echigo villages, transcending domain boundaries. No domain administration interfered with this village institution or fostered its development, yet redistribution practices remained an important part of village life until well after the Meiji land tax reforms of the 1870s. Detailed examination of actual practice in one village illuminates the care with which villagers approached the redistribution effort.

## Iwade Village

Iwade village has been the subject of considerable investigation—a testimony to the richness of the documents accumulated and saved by the Satō house. The *warichi* customs of the village have been a part of these studies but have not been the main subject of any.[2] Plate 1 shows Iwade's location west of the Shinano River, and a more detailed map (Figure 5.1) shows the immediate area of Iwade. The detailed area shown is about 10 kilometers square. Contour intervals are about 21 meters (69 feet), with highest elevations at about 900 meters.

Iwade village is located along the Kakizaki River in an area of relatively low mountains with good drainage. The Kakizaki is a small river, some 19.2 kilometers (just under 12 miles) in length from its headwaters to the Japan Sea. It drains just a small segment of Echigo Province, about 143 square kilometers (89 square miles). The Japanese government classification scheme ranks this as a second-class river based on its contemporary socioeconomic significance (rather than its physical attributes).[3] Nonetheless, flooding on at least limited scales was a historical reality. Although landslides occur in the Iwade region, modern techniques of mapping past landslides from aerial photography, satellite imagery, and the like show little impact on the village. Today, the risk of landslide is considered sufficiently low that the government has

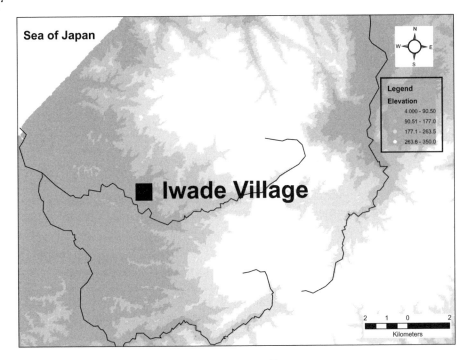

**Figure 5.1**   Iwade Village Area

not designated this area for special, preventative construction.[4] Despite these risks, the documents analyzed below do not mention multiple instances of flooding or landslide, an omission that suggests redistribution is taking place to accommodate changes in agricultural circumstances other than floods and landslides, that is, changes in soil fertility.

Iwade was a typical village for the Echigo area. It had an assessed value for land taxation of slightly more than 219 *koku*. In the late eighteenth century, it had a population of 140 people.[5] Throughout the Tokugawa era, the village found itself under the control of several different overlords. Several daimyo (Hori, Matsudaira, Sakai) alternated control for short periods before 1681, when the Bakufu took over. It controlled Iwade for most of the period to 1742. At that point, the Sakai, lords of Takada domain, took over control of the village until Meiji.[6]

During the time examined here, the Satō family consistently staffed the office of village headman. While the Satō had very large landholdings in other villages in the district and were thereby acquainted with the practices of other villages, there is no evidence that they brought outside practices to Iwade and no commoner officials such as *ojōya* could impose external practices on the community. Given the rotation of overlordship outlined above, even samurai officials likely had too limited a knowledge of local areas like Iwade to shape local institutions effectively even had they so desired.

As practiced in Iwade, *warichi* was strictly a village affair. The changes in practice described below came entirely from innovations originated by the participants themselves. Joint ownership practices in Iwade village can be traced back to the early eighteenth century, but we have no description of either its origins or the reasons villagers decided to employ these practices. Based on the long record of documents that describe the outcomes of redistributions, there is a pattern of conducting redistributions every ten years.[7]

## Documentary Evidence

The high quality of these documents notwithstanding, villagers left none that explicitly specify the principles on which redistributions were based. While documents attest directly to some chareacteristics, we must draw inferences regarding others. What we see overall is much more than a pro-forma process. Villagers reevaluated and remeasured village lands with care, tracking changes over time, despite the time and energy these efforts required.[8]

The Satō collection contains some thirty-five notebooks detailing the outcome of redistribution lot drawings. Overviews of them are presented in Tables 5.1 and 5.2. A large portion of them are completely intact. Other documents, not listed in these tables, are partial and undated. In addition to noting which shareholder got to manage what lands,

**Table 5.1**   Iwade Village *Warichi* Documents by Type of Land Recorded

|  | ALL | PADDY | DRY ONLY | PADDY, DRY ONLY | MT. ONLY | OTHER MT. | UNCLEAR |
|---|---|---|---|---|---|---|---|
| Arable (n = 23) | 2 | 9 | 4 | 8 | | | |
| Mountain (n = 12) | 2 | | | | 3 | 2 | 5 |

*Source:* Satō-ke monjo.

**Table 5.2**  Dated Documents on Redistribution of Arable Land

| Year | Satō Family Document No. | Title | No. of Wari | Complete? | Land Type |
|---|---|---|---|---|---|
| 1710 | 8006 | "Chiwari tanbun chō" | | complete | paddy, dry, residential |
| **1746** | **8013** | **"Denchi iritate genchō"** | **16** | **complete** | **paddy** |
| 1747 | 8372 | none | 7 | complete | dry field |
| 1756/1757 | 8009 | "Denpata chiwari chō" | 26 | complete | paddy, dry, residential |
| **1765** | **2554** | **"Jū nen ni tsuki kujikae Iwade-mura denchi jibun chō"** | **12** | **complete** | **paddy** |
| **1781** | **8014** | **"Denchiwari ku-jibiki chō"** | **27** | **complete** | **paddy, dry** |
| **1800** | **8373-1** | **"Denchi kujibiki aratame no koto"** | **31** | **complete** | **paddy, dry** |
| 1808 | 8373-2 | "Chiwari chō" | 4 | complete | paddy, dry |
| 1809 | 8374 | "Denpata chiwari gechō" | 33* | complete | paddy, dry |
| 1818 | 8012 | "Denpata chiwari kujikae chō" | 27 | complete | paddy, dry |
| 1818 | 8375 | "Denpata chiwari gechō" | 26 | complete | paddy, dry |
| 1840 | 8377 | "Nokori gojū-bu wari oboechō" | 2 | complete | paddy |
| 1841 | 8378 | "Nokori gojū-bu wari yachō" | 3 | complete | paddy |
| 1842 | 8379 | "Denpata chiwari yachō" | 12 | complete | paddy |
| 1848 | 8380 | "Suge no wari" | 2 | complete | dry |
| 1848 | 8381 | none | 2 | complete | dry |

*Note:* Bolded items are documents compiled for the Satō family's clearly private use.

* Takchara 50-*bu wari* (4 *wari*) not included.

the authors occasionally made additional comments about a piece of land, land rental, or special relationships that sometimes linked the outcomes of one drawing to two different sections of the village, and they sometimes noted land that was specially exempted from a redistribution. Such documents were prepared for internal village use or occasionally for the Satō family's own, strictly private, use, not for broad public dissemination. Consequently, even when explanatory comments are occasionally made, they are cryptic.

Of these notebooks, about two-thirds deal with the village's core farmland (*honden*), including all of the land classified as dry field or paddy (Table 5.1). A dozen notebooks deal with the allocation of parcels of mountain or hillside land. Although much of this mountain land may have been used as general common land (*iriai-chi*), some of it was cultivated and treated as arable.[9]

## General Outlines of Redistribution in Iwade Village

The total value of lands subject to periodic redistribution was a bit more than 193 *koku*, or about 88 percent of the village's assessed value. There were two exceptional areas. One was the lands belonging to a temple. The other comprised lands belonging to one particular family who, from the founding of the village, were treated as residents of a different village.[10]

Throughout the period covered by the documents, three broad categories of land were regularly included in redistributions; early in the period under review, there was also a fourth. Paddy and dry-field categories were typically redistributed together. Early in the eighteenth century, access to lands that the villagers designated as "residential land" (*yashikichi*), seedbed and garden plots, was also regularly incorporated in redistributions of paddy and dry field. The concern here was arable land and not houses themselves; agriculturally useful land, not buildings. As demonstrated below, there was some relationship between the redistribution of mountain lands and seedbed/garden plots so that, at least early in the period, some reallocation of mountain lands was likely to take place at the same time as paddy and dry fields. We do not know what happened to mountain lands after the redistribution of seedbed and garden plots lands ceased.

Iwade village *warichi* was of the proportional per share variety, that is, shareholders received access to all types and qualities of land in the village in proportion to the number of full and partial shares of cultivation rights that they held. No one could accumulate rights just in the best areas of the village. Further, it was possible to increase or lose the shares one managed. The large holdings of the Satō family are illustrative of precisely this characteristic. Within the village alone, they held rights to manage 30 *koku* in 1694 (Genroku 7), almost 93 *koku* in 1731 (Kyōhō 16), 121 *koku* in 1752 (Hōreki 2), and from 1798 (Kansei 10) to the Meiji Restoration (1868), they possessed 150 *koku*. In other words, from the mid- to late-Tokugawa years, they held between one-third and two-thirds of the arable land in the village.[11]

As the case of the Satō family makes abundantly clear, lands are not being redistributed to maintain equality of each participant's landholdings. As with all per share redistribution systems, Iwade's practice was not primarily a wealth-redistribuing mechanism. Since we do not have records for the start of *warichi* in Iwade village, we cannot know if this pattern of unequal land rights was present from the start. Even if the first participants all held rights to equal shares of land, inequality grew and fluctuated over time through the sale and purchase of shares.

Redistributable land in the village was divided into sixteen shares. While there is no indication of why this number was chosen—the number of shares does not appear to reflect the number of families in the village at the time—it was fixed by the end of the seventeenth century and remained constant thereafter.[12] If there were unequal holdings initially, the number of shares might have reflected the largest subdivision that would permit easy calculation of fractional holdings. In such a case, one or more families might have held rights to enough land to constitute $x + \frac{1}{16}$ share of village land, that is, $\frac{1}{16}$ share, or $1\frac{7}{16}$ share, or $5\frac{3}{16}$ shares. When villagers made their assessments of the different types and qualities of arable land in the village, each of the resulting parts of the village (what I call *wari*) was generally divided into sixteen sections for which lots would be drawn.

The land allotment documents themselves reflect the fundamental share structure, recording entries for each of the sixteen subsections of each *wari* in the village. For example, the entry

Hashi-muke, Yūeimon farmed [heretofore] paddy 5 fields
## Section 1                    [To] Yūeimon

begins with a designation of the place-name of the village area where the fields were located, followed by the name of the head of the shareholding family that had cultivated the land up to the time of the current redistribution, followed by the number of fields involved in the case of paddy or sometimes the land area if dry fields were involved. This was followed, in larger lettering, with the number of the section for which a lot was drawn, followed by the name of the new cultivator of that section. In the case cited here, the same family drew the lot for the section that they had cultivated previously—an outcome that was unusual but certainly possible in a random draw. In some cases, a share might be jointly held by multiple families, that is, families might have held fractional shares. In such a case, the name of the representative, the *kujioya*, of that share might be listed, or, if the number of people drawing for the share was as small as two or three, all names might be listed. Again, at the end of this process each shareholder had rights to cultivate a portfolio of lands comprising the same proportion of good, average, and poor quality paddy and dry field as every other shareholder.

The notebooks document a decennial pattern of implementation until the mid-nineteenth century. There are exceptions, but the dominant pattern is regular, especially for the first century of *warichi* operation. Let us look at the documentation and apparent variations from the decennial pattern a bit more closely (Table 5.3).

The listing in Table 5.3 differs from the years for which we have extant notebooks for arable lands (Table 5.2). The Meiwa 3 (1766), An'ei 4 (1775), and Kansei 3 (1791) redistributions are noted in the explanatory remarks in the Bunka 6 (1809) notebook. The Bunka 4 (1807) redistribution is noted in the Bunsei 1 (1818) notebook (in the allocation of the Takehara 50-*bu wari*).

Much of what appears to be a variation from this pattern (e.g., a nine- or eleven-year interval) is explained by the process of implementation in a given season. Evaluating the land, dividing it up into

**Table 5.3**    Years of Redistribution Implementation

| | |
|---|---|
| Hōei 7 (1710) | Bunka 4 (1807) |
| Enkyō 3–4 (1746–1747) | Bunka 5 (1808) |
| Hōreki 6–7 (1756–1757) | Bunka 6 (1809) |
| Meiwa 3 (1766) | Bunsei 1 (1818) |
| An'ei 4 (1775) | Tenpō 11 (1840) |
| Tenmei 1(1781) | Tenpō 13 (1842) |
| Kansei 3 (1791) | Ka'ei 1 (1848) |
| Kansei 12 (1800) | |

sections, and drawing lots all consumed a considerable amount of time: more than a month. Since these processes took time that could not be sacrificed during the agricultural peak season (spring to early fall), the redistributions typically were conducted over the last months of the old year and the early months of a new year. The Hōreki (1756–1757) notebook, for example, indicates that the process followed this schedule. While other documents may not say so explicitly, there are hints that they followed the same process. For example, a Meiwa 2 (1765) document (no. 2554) appears to suggest a break from this pattern, but a closer examination argues otherwise. The document is filled with notes about tenants and other comments that do not appear in most of the notebooks, suggesting that it was compiled as a private document, probably in preparation for the redistribution in Meiwa 3 (1766), a year for which we have clear evidence of a redistribution.

The evidence for a decennial pattern of implementation is strongest for the mid-eighteenth to early nineteenth century; the pattern begins to break down after the 1818 redistribution. The notebooks do not provide any direct indication of why the pattern changes at this time, nor is there any evidence of a significant economic, political, or social transformation in the Iwade area. The interval between the 1800 redistribution and the series of redistribution documents for the Bunka era—just eight years, not ten—suggests that there were pressures among village shareholders to reinvestigate and redistribute early. However, the documents for 1807–1809 present the first clear signs of trouble. Here we have activity over a three-year span. The implication is that the earlier efforts (1807–1808) led to unacceptable results in the eyes of a significant portion of the participants. Consequently, a further reallocation was undertaken in 1809. We have documentation for a redistribution nine years after the 1809 reevaluation (in 1818) but no evidence of a redistribution in 1828 or 1838—the pattern one would expect if there were a reversion to the ten-year interval. The next redistribution (1840) was conducted only on paddy, and it was followed within two years (1842) by another, again focusing only on paddy. These incomplete redistributions all suggest dissent among Iwade's shareholders. One might speculate that the Tempō agricultural crisis introduced new stresses on the decision-making process, but there are other possibilities. No contemporary documentation explains this break from the ten-year interval, but other evidence from disputes confirms difficulties that likely disrupted regular decennial redistributions.

That change was afoot is neither surprising nor unusual. Even in the mid-eighteenth century, participants made a shift in allocation of "residential garden land" (*yashikichi*). The seedbed and garden plots land category came to be limited to accounting for the adequacy of land near a house that could be used for vegetable gardening or seedbeds. In other words, the core garden/seedbed plots near residences were not rotated, only those that were allocated to compensate for an over- or underallotment of such lands. Although documents contain no explicit standard, the shareholders who did not have access to their full share were provided with the right to draw additional shares of mountain land to make up for the shortfall. Thus, drawings for seedbeds and garden plot *wari* were noted in the arable land *warichi* registers up through the 1756–1757 iteration, but thereafter none of the notebooks have any sections designated as residential/garden plots, just paddy, dry field, or mountain land. Shareholders decoupled reallocation of mountain lands and arable lands. The shift is indicative of a willingness to reconsider redistribution principles and to alter them as perceptions and feelings changed over time.

Pre–Meiji Restoration documentation gives no hint of what underlay the emerging mid-nineteenth-century pattern of irregular redistributions, However, in early Meiji, just before the implementation of the Land Tax Reform, there is an indication that landlords like the Satō were unhappy with the process. A protest erupted in Iwade village in 1870 (Meiji 3) that included a demand to reimplement *warichi* redistributions.[13] In the course of the disturbance, officials from the Meiji government's Office of Civil Administration (Minseikyoku) investigated. In replying to their inquiries, village officials (who were also landlords) noted that in the year when a redistribution was upcoming, tenants and small holders did not assiduously maintain their fields, and, as a result, their fertility was compromised. This complaint suggests that landlords had grown wary of redistribution in the mid-nineteenth century, fearing the impact of free rider problems, and refused to go along with a periodic schedule of redistributions. They were only willing to implement partial redistributions in specific sections of the village when a landslide or flood introduced some significant change in the land or when some issue of equity in land valuations arose among cultivators. (Readers will recall that this was the practice throughout much of the Tokugawa era in Kaga domain and that it largely avoided free rider issues because the decision to implement could not be preplanned or calculated as with periodic redistributions.) Ignoring the landlords' explanations, the Civil Administration deferred to local custom and asked that a redistribution be implemented, a request that the village followed. There was no validation of landlord claims of free riders.

The landlords' complaints raise a significant question as to why a very large (more than two-thirds of Iwade village land) landlord like the Satō family would participate in redistribution rather than just acquire all of the best land for themselves. One rationale, suggested below in connection with disputes with tenants (see Chapter 8), is that it was demanded by tenants on whom the Satō depended to cultivate most of the land they managed. Even within the holdings of a single large landlord there would be significant variation in land quality. To reduce the possibility of failure of tenant households, and therefore a loss of rent revenue, assuring all tenants access to a comparable portfolio of lands served the interest of the landlord as well as tenants. Still another possibility lay in a desire to protect and maintain the viability of other landholding families who, as part of the village upper crust, shared with the Satō all of the burdens of running the village. In an agricultural environment heavily dependent on irrigation and the collaborative efforts essential to maintain and run the system in a limited (but growing) cash economy, large landlord participation in joint ownership created a more broadly based class of cultivators with a vested interest in the smooth operation of essential services to agriculture. Regardless of the motivation, the Satō never bought up all of the land in the village, apparently content with their dominant position in the existing tenurial regime.

## Dividing the Land: Creating *Wari* Divisions

The land divisions described in Iwade documents display considerable variation in the way in which paddy, dry field, seedbed/garden plots, and mountain lands were classified by the villagers over time. Tables 5.1, 5.4, and 5.5 (below) indicate that shareholding villagers divided all arable and residential lands into as few as twenty-six sections *(wari)* and as many as thirty-three, including allocation of residential/garden land. For mountain lands, the variation ranges from a low of ten to a high of fifteen *wari*. There is much change from notebook to notebook and therefore from redistribution to redistribution; there are very few instances of complete notebooks in which the same number of *wari* subdivisions are recorded in more than one document. Tables 5.4 and 5.5 include notebooks that represent full surveys of either village arable or mountain lands. These notebooks record data for more than sixteen

**Table 5.4**　Variation in Number of Mountain Land *Wari* Found in
Complete Redistribution Documents (Complete *Wari* Only)

| No. of Documents | No. of Subdivisions (*Wari*) Recorded | Land Type |
| --- | --- | --- |
| 1 (?) | 15 | Mountain Dry Field/Mountain |
| 1 | 11 | Mountain Dry Field/Mountain |
| 1 | 10 | Mountain only |
| 1 | 6 | Mountain only |
| 1 | 5 | Mountain only |
| 1 | 3 | Mountain only |
| 2 | 2 | Mountain Dry Field only? |

*Note:* The document noted in the first data row with the question mark may
not be complete; within the listings of *wari*, there are two *wari* that appear to
be missing a single subsection.

*Sources:* Satō-ke monjo, documents 8010, 9011, 8376, 8018 (1–8), 8373-3

sections *(wari)* in the village for arable lands; for mountain land, they
record data for more than six. Notebooks other than these are either
clearly incomplete (with pages obviously missing) or likely incomplete,
and therefore they are useless to investigate variation in the way that
shareholders subdivided the village into sections.

Variation in the way in which the villagers subdivided arable lands
indicates that they approached a new redistribution unrestricted by any
fixed tradition as to the number of *wari* they should employ. They en-
tertained new ways of thinking about village arable and mountain lands
at each iteration. Villagers subdivided or recombined *wari* divisions.
Land reclamation and significant land loss in Iwade was uncommon
during the period covered by the documents, so increased or decreased
arable acreage does not account for this variation. While the rationale
that underlay each pattern of dividing up the village land may not be
apparent, villagers genuinely revisited the question of what constituted
a section of uniformly comparable quality land.

What land was to be included, how often to redistribute, and how
to divide up the lands in the village into sections were all subject to
reconsideration. At the heart of all such issues lay the proper evaluation of the quality of village lands, dividing them into *wari* of uniform
quality, and marking their boundaries. More detailed analysis of this
process over time is revealing.

**Table 5.5** Variation in Number of *Honden Wari* Found in
Complete Redistribution Documents

| No. of Documents | No. of *Wari* | Land Type | Remarks |
|---|---|---|---|
| 1 | 33 | Paddy/Dry | |
| 2 | 31 | Paddy, Dry Residential, or only Paddy and Dry Field | |
| 2 | 27 | Paddy, Dry | |
| 2 | 26 | Paddy, Dry Residential, or Paddy and Dry Field | |
| 1 | 16 | Paddy | |
| 1 | 14 | Paddy | Part of document missing |
| 1 | 13 | Paddy/Dry | First part of document missing |
| 4 | 12 | Paddy | |
| 1 | 11 | Paddy | Part of document apparently missing; within the listed *wari* there are two that appear to be missing a single subsection |
| 1 | 7 | Dry | |
| 1 | 4 | Paddy/Dry | |
| 2 | 3 | Paddy, Paddy/Dry | |
| 3 | 2 | Dry (2 documents), Paddy (1 document) | |
| 1 | 1 | Dry | |

*Sources:* Documents listed in Table 5.2.

Table 5.6 records these data for dry field and paddy through the first half of the eighteenth century. For each year in which a particular *wari* was recorded in the documents, its area in *bu* and the number of shares into which it was divided are noted. The left-hand column records the name of the *wari* as recorded in the redistribution notebook. For each of the next three columns, the size of the *wari* subsection and the number of shares into which it was divided are shown for three redistributions for which we have notebooks, 1710, 1746, and

**Table 5.6** Changes in *Wari* Names, Area, and Number of Shares per *Wari*,
1710–1757

| | Subsection Size *(bu)* / No. of Shares | | |
|---|---|---|---|
| W'RI NAME | 1710[1] | 1746 | 1756/1757 |
| 60-*bu wari*; paddy[2] | 60/16 shares | | |
| 230-*bu wari*; paddy | 230/16 shares | | |
| 3 *tan* with dry field; paddy | 300/16 shares | | |
| Matakita 350-*bu wari* (2?); paddy | 350/16 shares | | |
| 100-*bu wari*; paddy[3] | | | 100[4]/16 shares |
| Machiura *wari* (2);[5] paddy | | 360/16 shares | |
| 80-*bu wari*; paddy | | | 80/16 shares |
| **50-*bu wari*; dry[6]** | <u>50/9 shares</u> | | |
| **Yashiki-*bu*; dry** | <u>?/9 shares</u> | | 240/24 shares |
| **Dry field 50-*bu wari*; dry** | <u>50/12 shares</u> | | |
| **50-*bu wari*; dry** | 50/16 shares | | |
| **60-*bu wari* (2); newly re-claimed dry** | 60/16 shares | | |
| **100-*bu wari* (3); dry?[7]** | 100/16 shares | | |
| **Nagatoro Kawahara *wari*; dry?** | <u>?/4 shares</u> | | |
| **Nakama Dry Field; dry** | | | ?/6 shares |
| **Kawahara Dry Field; dry** | | | 30/16 shares |

*Note:* The *wari* names that appear in this chart do not appear in any later documents. Each column headed by a year shows the number of *bu* for the *wari* and the number of shares into which it was divided. Bold entries are dry-field *wari*; underlined numbers highlight those *wari* that were divided into fewer than sixteen shares. *Wari* sizes represent the size of each share contained, not the size of the whole section being allocated. Thus, 60 *bu* = about 198 square meters, 230 *bu* = about 760 square meters, and so on.

[1] *Wari* without names are relatively common in these three documents and more common in paddy than in dry field.

[2] The number is not clearly recorded, but the number of fields is. In general, the number of fields is not recorded for dry field.

[3] The number of *wari* is conjectural.

[4] The document records the following note: "Shares are drawn together with those in the 80-*bu wari*. This was also done for lots drawn for Gotanda and Onidani *wari*."

[5] Later, likely the same as "Uraoki *wari*."

[6] Generally, the number of fields is recorded for paddy and not recorded for dry field.

[7] Generally, the number of fields is recorded for paddy, but that data is not recorded for this *wari*.

*Sources:* Satō-ke monjo, documents 8006, 8372, 8009.

1756–1757. Blanks in the cells indicate that the *wari* was not recorded in the document for that particular year. For example, the 60-*bu wari* was recorded only in the 1710 notebook, and the subsections of the *wari* were 60 *bu* in size. The subsection was divided into sixteen shares.

Since there are no maps associated with the documents from which these data are drawn, it is hard to tell what was being done with *wari* listed in the earliest of these documents. Some *wari* disappear, and others appear over time. One cannot discern conclusively for each *wari* whether the same section simply had been labeled with a different name or if the section was split up or combined to make radically new and different divisions within the village.

Nonetheless, while there are changes evident in the early eighteenth century, the divisions for *honden* paddy that we see in 1756–1757 remain fixed throughout the remainder of the Edo period. So, too, do the sizes of these *wari*. These sections of the village thus appear to have stabilized by the mid-eighteenth century.

In other parts of the village, however, the paddy field *wari* are less stable (Table 5.7). The two Shimonishi *wari* first appear in the two mid-eighteenth-century redistributions, and the *wari* sizes were presumably the areas mentioned in their names. In the 1781 redistribution, the size of these *wari* changed. Thereafter, their size remained stable. Similarly, the sections within the Sawada *wari* changed from 70 *bu* to 60 *bu* in 1800 and then back to 70 *bu* thereafter.

While the shifts in paddy seem concentrated early in the period for which we have documents, the situation is more complicated for dry field. Like paddy, dry-field *wari* names that appeared in the 1710 notebooks almost completely disappear thereafter (Tables 5.6 and 5.8). Three of the new *wari* in 1746 remain stable in terms of area to the mid-nineteenth century. (There are changes in the number of shares into which some are divided, however: for example, Asabatake *wari*, 100-*bu wari*, and Shinden *wari*.) As Table 5.8 shows, there are nine *wari* that only appear once or twice in the entire series of documents. If we add to that number those that only appear three times (in about one-quarter of the redistributions), we have fourteen different cases of transient dry field *wari*. Within these, the 100-*bu wari* was, until 1756, divided into three but appears to become just one *wari* in 1781. Takehara 50-*bu wari*, originally divided into four, was reorganized in 1809 into two 100-*bu wari*.

In area, too, the changes are much more evident in dry-field sections of the village than in paddy. Table 5.8 shows that Shimakura

**Table 5.7** Changes in Paddy *Wari*, Subsection Size, and Number of Shares, 1746–1840

| | 1746–1747 | 1756–1757 | 1781 | 1800 | 1809 | 1818 KUJIKE CHŌ | 1818 GECHŌ | 1840A | 1840B | 1842 |
|---|---|---|---|---|---|---|---|---|---|---|
| Sawada | 4 *wari* 70/16 shares[1] | 1 *wari* 70²/16 shares | 1 *wari* 70/16 shares | 1 *wari* 60/? | 1 *wari* 70/16 shares | 1 *wari* 70/16 shares | | | | 1 *wari* 70/16 shares |
| Shimo Nishida 250 *bu* | | | 125/16 shares | 125/? | 125/4 shares | | 125/4 shares | 125/4 shares | 125/4 shares | |
| Shimo Nishida 100 *bu* | | | 100/4 shares | 100/? | 100/4 shares | | 100/4 shares | 100/4 shares | 100/4 shares | |
| Takawahara³ | | | 50/? | | | | | | | |
| Ueda | | | | | | | | ?/16 shares | | |

*Note:* Except where there are no data, the cells for each *wari* display the number of sections included, the size of the *wari* in *bu*, and the number of shares into which each *wari* was divided. Share numbers other than 16 are underlined. *Wari* sizes represent the area each share contained, not the size of the whole section being allocated, so 70 *bu* = about 231 square meters, 250 *bu* = about 827 square meters.

[1] For the 1746 redistribution in the same document, there is also another *wari* with this name, but the size of the *wari* is listed as 17 *bu*.

[2] The document notes that drawing shares is *ippon kuji*, probably indicating that one lot drawn simultaneously selects sections of this *wari* plus the similarly named but smaller *wari* identified in note 1.

[3] The number of subdivisions in the *wari* is not recorded. This suggests that these are the document author's allocations only.

*Sources:* Satō-ke monjo, documents 8009, 8012, 8372, 8373-1, 8373-2, 8374, 8375, 8377, 8378, 8379.

**Table 5.8** Dry-Field *Wari* Changes, 1710–1848

| | 1710 | 1746–1747 | 1756–1757 | 1781 | 1800 | 1809 | 1818 KUJIKE CHŌ | 1818 GECHŌ | 1840A | 1840B | 1848 |
|---|---|---|---|---|---|---|---|---|---|---|---|
| Shimakura-Kawahara | ?/4 | | | | 20/16 | 20/16 | 30/16 | 30/16[1] | | | |
| Asabatake (2 *wari*) | | 30/16 | 30/16 | | 30/? | 30/17&15 | | 30/17&15 | | | |
| Hyaku-bu-hatake (3 *wari*) | | 100/16 | 100/16 | 3→1 *wari* 100/? | | | | | | | |
| Shinden-batake | | 60/32 | 60/20 | | 60/? | 60/20 | 60/20 | 60/20[2] | | | |
| Shimonishi | | | | 60/4 | 40/? | 14/16 | | | | | |
| Senzei-batake | | | | 20/16 | | | | | | | |
| Takahara (4 *wari*) | | | | 50/? | | | | | | | |
| Kaminishi (2 *wari*) | | | | 30/? | | | | | | | |
| Takehara (4 *wari*) | | | | | 50/? | 50/16 | | | | | |
| Takehara (2 *wari*) | | | | | | 100/16 | 100/16 | 100/16 | | | |
| Takehara | | | | | | (no #2 *wari*) 50/24 | 50/24[3] | 50/24[4] | | | |

*(continued on next page)*

**Table 5.8** (continued)

| | 1710 | 1746–1747 | 1756–1757 | 1781 | 1800 | 1809 | 1818 KUJIK'E CHŌ | 1818 GECHŌ | 1840A | 1840B | 1848 |
|---|---|---|---|---|---|---|---|---|---|---|---|
| Nagatoro Shin-*wari* | | | | | | 33/16[5] | 45/16 | 45/16[6] | | | |
| Nagatoro-Sugeno | | | | | | 50/16 | 50/16 | | 43.5/16[7] | 50/16[8] | nominally 60; 67.625/16[9] |
| Nagatoro-Koshiba | | | | | | ?/17&16 | | | | | |
| Nagatoro-Sassabu-chi-Kayano | | | | | | 20/15 | 20/16[10] | | | | |
| Monzenseki-Tsumekayano | | | | | | 50/16 | | | | | 74.8/16[11] |
| Ozeki-Kawahara-Sugeno | | | | | | | 50.5/16 | | | | |
| Yamasawa-Yoshino[12] | | | | | | 30/6 | | | | | |

*Note:* The cells for each *wari* show the number of sections included, the size of the *wari* in *bu*, and the number of shares into which each *wari* was divided. Share numbers other than 16 are underlined.

[1] "Shimakura 30-*bu wari*."

[2] "60-*bu wari*." Within this *wari*, more detailed explanations indicate that parts of it were either "previously reclaimed land" (*mae shinden*) or relatively recently developed reclaimed land (*naka shinden*).

3 "Takehara 50 *bu* 3 *wari*."

4 "Takehara 50 *bu* 3 *wari*."

5 "Nagatoro Shin-wari Yanagihara Shinbatake 33-*bu wari*."

6 "Nagatoro-batake 45-*bu wari*."

7 "Nagatoro-tofu *wari*."

8 "Nagatoro-tofu *wari*."

9 Nominal wari size is 60 *bu*; listed area was based on actual measurement.

10 Nagatoro-Sugeno 20-*bu wari*, Sassabuchi.

11 Actual measurement.

12 Data in documents noted here appear to be incomplete.

*Sources:* Satō-ke monjo, documents 8006, 8009, 8012, 8372, 8373-1, 8373-2, 8374, 8375, 8377, 8378, 8379, 8380.

Kawahara *wari*, Shimonishi *wari*, Nagatoro Shin *wari*, Nagatoro Suge-no *wari*, and Monzenseki Tsumeno *wari* all changed in size. All of this took place in addition to the 1809 reorganization of the Takehara *wari* and the creation of a new Takehara 50-*bu wari*. (The considerable attention paid to dry fields evident here will be discussed below and in a different context in Chapter 7, which suggests that shareholders placed special value on them.)

In sum, while organization of paddy sections of the village was overwhelmingly completed by the mid-eighteenth century, organization and reorganization of dry-field sections of Iwade continued long after that and suggests that shareholders felt a need for continued reassessment of these sections of arable land.

In addition to the paddy and dry-field *wari* discussed so far, some lands did not fit into a standard *wari*. These were labeled *zanbu*, or "remaining *bu*." Their area was very limited, and they appear to constitute small, scattered fields. Their treatment varies somewhat over time (Table 5.9). The Enkyō (1746–1747) redistribution allocated dry-field *zanbu* by drawing lots; however, in the Hōreki (1756–1757) redistribution, the two *zanbu wari* were not allocated by lot, but assigned to individuals in compensation for the fact that they were not allocated a full share of land elsewhere (e.g., for *yashiki* seedbed/garden plots land). Before the Tenmei era (from 1781), the amount of *zanbu* fluctuated, but thereafter it was very stable.

## Measuring *Wari*

In Western European tradition, assessment of the quality and value of land is typically treated as a separate process from measurement of land area and the setting of boundaries; in Japan, the two processes tend to be confounded in the early modern era.[14] This can be observed in the *warichi* practices of Iwade Village (and elsewhere in Echigo). In the *warichi* villages like Iwade, the confounding was a simplification of the process that divided village arable into sections of comparable overall productivity and quality, including exposure to beneficial and harmful characteristics of the natural environment. Land survey practices throughout Japan are replete with officially recognized adjustments to actual measurements to reflect deviations from per hectare yield assigned to lands. For example, a 20 percent adjustment for strips of land in shadow was standard. So Iwade villagers were not exceptional.[15]

**Table 5.9**  Changes in Lands outside of *Wari*, 1746–1842

| | 1746-1747 | 1756-1757 | 1781 | 1800 | 1809 | 1818 KUJIK'E CHŌ | 1818 GECHŌ | 1840A | 1840B | 1842 |
|---|---|---|---|---|---|---|---|---|---|---|
| Leftover | | | | 50/? | 2 *wari* 50/16[1] | 2 *wari* 50/16 | 2 *wari* 50/16 | 2 *wari* 50/16 | 3 *wari* 50/16 | |
| Leftover paddy/dry | | various/9 places[2] | | | | | | | | |
| Kawahara leftover | | | | | | | 47/16 | | | |
| **Leftover Dry Field** | ?³/4 | various/21 places | 30/16 | 30/? | 30/16 | | 30/16[4] | | | |
| **Leftover Dry Field** | | | | | | 40/16 | 40/16[5] | | | |

*Note:* The cells for each year indicate no. of *bu*/ no. of sections in the *wari*. Bold *wari* are dry field; underlined numbers are the number of sections in the *wari* when that number is not 16.

[1] From 1808 (Bunka 5) divided into two *wari* of 50 *bu*, sixteen shares each.
[2] Scattered in various places rather than in a single, contiguous *wari*.
[3] Total 520 *bu*; about 70–75 *bu* per share.
[4] "Nokori-batake 30-*bu wari*."
[5] "Kawahara Zanbu-batake."

*Sources:* Satō-ke monjo, documents 8009, 8012, 8013, 8373-1, 8374, 8375, 8377, 8379.

Although the valuation and area measurement processes were not distinct, actual measurement of land was a significant part of the redistribution process. Even when abbreviated, measurement and assessment were very labor-intensive tasks and consumed more of villagers' efforts during redistribution than any other process. Yet the evidence from the Iwade village notebooks suggests frequent remeasurement of at least some parts of the village.

Changes in the size of *wari* suggest some limited actual measurement during redistribution. As a rule, while the redistribution notebooks record the number of fields for paddy, the actual area typically is not recorded; however, occasionally there is a recording of area in *bu*. For example, in the Hōei 7 (1710) *Tanbunchō*, the seventh *wari* (unnamed) entry provides one of several examples of actual measurement: sections 4 to 7 are noted as being (respectively) 360 *bu*, 300 *bu*, 350 *bu*, and 390 *bu*, with some of these clearly recorded as having more or less area than they would have possessed in order to maintain the principle of parity in the valuation of specific sections. In the 1740 (Enkyō) privately compiled notebook, we find notation of the area (in *bu*) of paddy let out to tenants. Among several examples in the 1809 redistribution (no. 8374), in an explanation of Uranaka Number 2 *wari*, section 7, we find, "In addition to these thirty fields, 20 *bu* in Nagatoro." While there are occasional instances of area measurements of more than 100 *bu* after 1740, the 1809 notebook is representative of the overwhelming number of cases: the area of actual measurement was typically 30 *bu* or less.[16] This suggests that actual measurement was conducted only when deemed essential and then only partially, usually on relatively small fields.

The impression that villagers conserved their energies and measured fields only on the basis of special need is strengthened further by the fact that measurements were not necessarily repeated in a *wari* subdivision in subsequent redistributions. In this regard, villagers saved themselves some labor. Iwade villagers were not unusual. Other communities eliminated remeasurement where they could as well. Given that a redistribution that included remeasurement might take six weeks or more, such conservation of effort is understandable.[17]

In addition, these paddy measurements were rough, not precise. In all of the tables showing arable *wari* size, the area noted is approximate.[18] For example, in the case of arable land (as opposed to mountain lands), in cases in which the unit of measure is a half *tan* or smaller, areas are measured in 5-*bu* intervals, also suggesting relatively rough measure. Shareholders focused their concern on overall output of the land in each section of a given *wari* rather than precise calculation of the area involved (all exceptional cases involved small areas).

These rough and imprecise measures suggest eyeball estimates, the outcome of discussion and establishing shareholder consensus on a given area's output rather than actual use of a measuring line.[19] Iwade shareholders were not concerned with small differences in area. Instead, they paid attention to only those instances in which there was a sufficient shortfall in overall output of a plot that they felt obliged to compensate by allocating supplementary lands in another part of the village. They even gave a name to this condition of insufficiency in *warichi* notebooks: *kabusoku*, literally "insufficient share."

What of dry fields? These are widely considered less valuable than paddy, and less central to agriculture, so one might expect they were a less intense focus for measurement. Like paddy, records for dry-field *wari* typically do not note the size of shares within the *wari*, but when they do, size is recorded in whole *bu*, suggesting again an approximate rather than precise measurement, albeit more precise than for paddy. Again, dry fields drew more attention than paddy.

Shareholders treated seedbed and garden land in a very different manner, however, measuring quite precisely and using measurement units as small as a Japanese foot (*shaku*, about eleven inches). The 1740 *Tanbunchō* uses units of length only *(ken* [about two yards] and *shaku),* not a unit of area such as a *bu* or a *tan,* and is precise to tenths of a *shaku.* The same is true for the 1756 redistribution notebook. These documents even calculate the degree of allotment shortfall *(busoku)* with the same degree of precision.

The precise measures suggest the special value of seedbed and garden plot lands to shareholders. Only in the case of small parcels not included in a standard, well-organized *wari (zanbu)* did they employ this degree of precision to measure other arable lands (in the case of *zanbu* we see measures as small as tenths of *bu*).

The emphasis on precision of measurement should not obscure one very important point: whether they measured all of the arable or just a part of it during a redistribution, participants stressed parity in the output of the land in each section within a *wari* rather than equal area. This emphasis is noted throughout the redistribution notebooks in remarks detailing compensation for the shortcomings of a section in a particular *wari* as well as other comments on the quality of land in a section: "in addition, with grasslands," "with dry field," and similar comments are scattered throughout these documents in describing *wari* that are designated as paddy. The presence of such nonpaddy lands within a paddy section meant that crops with output as high as rice could not uniformly be grown there. In such cases, even if the section was larger than others in a *wari*, its overall output was deemed, by agreement of all shareholders, to be equal to the other subdivisions of that *wari*, and it would not matter if one drew a lot for one section of land rather than another within it. Furthermore, random drawing of lots generally assured that no one could count on consistent access to a larger or better field, meaning that even if there were a small difference, only fate determined who received benefits or paid a price in a given reallotment.[20]

In sum, large changes in area are infrequent and usually measured. Even when there was a notation "not measured" *(sao irazu)*, shareholders discussed and reached an agreement about the changes in the size of each paddy and dry field, or conversion of lands from one to the other, and decided whether to measure and reorganize the land in each *wari*. On some occasions they redrew and remeasured sections; on others, they simply used the old divisions without remeasurement. Measurement of "remaing *bu*" *(zanbu)* and seedbed/garden plot lands was, by contrast, quite precise and consistent, perhaps because seedbed and garden plot lands were considered especially important, on the one hand, and the plots of "remaining *bu*," on the other hand, were small and easily measured.[21]

Villagers' failure to use actual measurements consistently in allocating all lands and the varying size of a share of land from one *wari* to another was embodied in the label they gave to each share, *kenmae*. Iwade village's use of this term reflected common parlance in the Echigo region. Translating it as "share"—devoid of a sense of specific land area or value—reflects the perspective of villagers. While villagers' use of *kenmae* to calculate their access to land may strike us as vague and difficult to grasp, villagers had a fair, consensus-based understanding of the portfolio of land rights one was getting when a family received (or transferred) a full or partial share. Land area was a part of that assessment but was tempered by a cognizance of the land's overall output so that villagers stressed an estimate of land value that combined and conflated area and per hectare yield. That overall assessment of land quality was paramount in their thinking.

## Variant Patterns of Subdividing *Wari*

Since, in principle, the cultivation rights to village land were divided into sixteen shares, one would expect that villagers allocated lands in each *wari* to individual holders by drawing lots for each of sixteen subsections. While this was the rule, in practice villagers sometimes employed a very different pattern of subdivision to deal with special circumstances. In the notebooks that deal with paddy and dry field, about half clearly list all shareholders, but the biggest exceptions appear in the handling of mountain lands.

Mountain lands could be the source of kindling, household building materials, and similar supplies, but they were also areas where dry-field crops could be planted. Nonetheless, only about one-fourth of the mountain land documents clearly list all shareholders (Tables 5.10, 5.11). In Table 5.10, the six notebooks for mountain lands that do not list all sixteen shares respectively list eight, six, and one share. In other words, not every shareholder was being allocated land in each section of the village's mountain land. This may simply suggest that villagers

**Table 5.10**   Documents Recording Shareholders

| ALL SHAREHOLDERS LISTED? | ARABLE | MOUNTAIN | TOTAL |
|---|---|---|---|
| All | 12 | 3 | 15 |
| Other than missing parts of Iwade, all | 3 | 0 | 3 |
| Apparently all | 1 | 0 | 1 |
| Clearly not all | 7 | 3 | 1 |
| Unclear (unlabeled maps) | 0 | 6 | 6 |

*Source:* For paddy and dry field, documents listed in Table 5.2; for mountain, documents 8010, 8011, 8376, 8018 (1–8), and 8373-3.

**Table 5.11**   Documents That Fail to Include All Shareholders

| | PADDY | DRY | PADDY & DRY | MOUNTAIN |
|---|---|---|---|---|
| Cultivators only listed | 4 | 0 | 0 | 0 |
| Revised reallocations only | 0 | 0 | 3 | 0 |
| Reason unclear | 0 | 0 | 0 | 6 |

*Source:* For paddy and dry field, documents listed in Table 5.2; for mountain, documents 8010, 8011, 8376, 8018 (1–8), and 8373-3.

felt that the variations in the quality of mountain lands were just too small to worry about, and allocations took place across some sections (with people drawing lots in only one of the sections that were explicitly paired) rather than just within them.

The following cases illustrate these unusual patterns in *wari* subdivision. In the 1768 redistribution of mountain lands (no. 8010), four *wari* were divided into fourteen subsections (Sukezawa, Hisaueimonsawa, Shironosawa, and Dōchūsawa). A thirteenth, unnamed *wari* was divided into eight sections. In the examples divided into fourteen sections the following note appears: "However, there are exclusions because two sections of mountain land were drawn that were over the allotment." In this case, because lands drawn in another section of mountain land were excessively large, two shares were not drawn in these particular sections. This process was deemed easier than subdividing and remeasuring this section of mountain land—at least relative to its economic value to the villagers. However, for the *wari* divided into eight sections, there is no such explanation. Similarly, in another, undated document (no. 8373-3), the second *wari* recorded (no name is attached) was divided into only seven sections, and Kaya *wari* was divided into seventeen sections. No explanation was provided.

There are examples of *wari* composed of fewer than sixteen shares for arable land, too.[22] In the incomplete data for 1710, two *wari* were divided into nine subsections, one into four, and two into twelve. In the same manner, in the 1756–1757 redistribution notebook, two *wari* were divided into six subsections. In the 1809 implementation, one *wari*, Senshō Sawada 70/Yamasawa Yoshino 30-*bu wari*, was divided into six subsections.[23] In four other notebooks, most sections were divided into sixteen subdivisions, but there were others in which there was either no subdivision, two subdivisions, or four subdivisions.

Finally, the number of people participating could also differ from *wari* to *wari*. The number of participants drawing lots from some sections broke with the sixteen-share rule and included only some of the participants. For example, in 1710, Hatake 50-*bu wari* formally comprised "nine-man groups," but actually there were only six people participating; in the same redistribution, the seventeenth *wari* comprised four people and the so-called Sukesawa nine-man group, in fact, had only eight people participating; and the Hatake 50-*bu wari* had only seven participants.

Other variations from standard practice involved a number of participants larger than those typically listed as drawing lots in an individual *wari*. For example, in the "Remaining *bu*" *wari* (*zanbu wari*, Hōreki, 1756–1757) noted above, there were six individuals, and in the Nakama-batake *wari* there were six participants, but in the list of individuals participating in the complete redistribution, we find ten people. Among the 1809 (Bunka 6) *wari* data, we find nineteen names listed, somewhat larger than the sixteen sections into which each *wari* was divided.[24]

Whatever motivated this special treatment, it was limited to a small number of cases and was of marginal import relative to the total productive value of arable land in the village. The rationale for this kind of special treatment of some *wari* is not clear. Nothing in the documentation suggests that there is a tradeoff of rights in one section for another or that there was compensation for an overallotment of seedbed/garden plots or other land. In sum, regardless of the rationale, unexplained deviations from principle were limited.

In looking at the difference in numbers of participants who shared each *wari* (both in arable and mountain sections of the village), there is also a suggestion of inequity in allotments. Some of it can be explained as relating to seedbed and garden plot land allotment; but such instances cannot explain all of the apparent inequity. While I cannot offer an explanation at present, from the earliest Iwade *warichi* notebooks there are hints of a departure from principles of equity that presumably underlay the system. This inequity continued in a variety of guises through the end of the Tokugawa era.

## Conclusions

The preceding analysis of Iwade village joint ownership reveals a practice that involved substantial shareholder discussion. Based on negotiations, the resultant practice deviated in some respects from that one would expect to see based on standard descriptions of *warichi*. It was a malleable institution, not ossified. Villagers took pains to carefully reassess their fields as part of the reallocation of cultivation rights even though they employed approaches that in modern times would be considered inexact. Most care was directed to management of dry fields, not paddy.

In the eighteenth century, redistributions took place on a fairly regular basis, about every ten years, but in the nineteenth century this pattern began to dissipate. By the mid-nineteenth century, documents show an increase in partial redistributions. While there is no explanation in the documents, one might speculate that it was the result of dissatisfaction and shareholder protest.

By the middle of the eighteenth century, the core paddy *wari* were established. Only minor changes were made throughout the rest of the era. However, noncore paddy, dry field, seedbed/garden plots, and lands not included in established *wari* were treated with considerable variation over the entire time for which records survive. Even before the mid-eighteenth century, the Hō'ei (1710) and Hōreki (1756–1757) redistributions reveal few changes in paddy, the fields typically seen as most central to agriculture. With the exception of the notebooks between the Hō'ei and Hōreki eras of the early eighteenth century, much of the evolving redistributional practice in the eighteenth and nineteenth centuries focused on dry field, not rice paddy (for example, variation in the number of *wari*, changes in area, appearance of new *wari*, and so on).

Further, while villagers actually measured lands in preparing for some redistributions, they skipped measurements or took shortcuts in others. They saw full remeasurement and reclassification of all arable land as burdensome work. They avoided it when possible, when they sensed no dramatic change in village conditions. Then they simply redrew lots for shares in sections defined by the previous redistribution.

Rather than careful measures of land area, villagers stressed estimates of the overall yield of each share. In other words, the lack of measurement did not signify poor understanding of the value of land in each *kenmae* share. From the villagers' perspective, the area of land was one consideration determining a share's overall yield. In small areas of land or where villagers placed a high value on land, as in the treatment of seedbeds and garden plots, shareholders actually measured land area with some precision.

In principle, each *wari* was divided into the same number of subsections as there were *kenmae* shares (sixteen), and cultivation rights were distributed to each household according to the number of *kenmae* each possessed. However, in Iwade village some *wari* were not divided into sixteen shares; not all shares were represented, and not all shareholders participated. These *wari* were not located in the core lands but nonetheless appear to represent some inequity in the handling of shares.

How redistribution of mountain land fit with redistribution of arable and seedbed/garden plot land is uncertain. From time to time, mountain dry field was included in arable *wari*, but from before the Hō'ei era it was typically treated separately from arable sections of the village. In addition, since the per hectare yields on both mountain and dry fields were probably low, villagers may have made the judgment that tolerating inequities in these parts of the village was better than investing the time it would take to eliminate them. In discussions with participants who lived where both *warichi* and swidden *(yakihata)* continued into the mid-1960s to early 1970s, such an attitude was not at all uncommon; however, this explanation is speculation on my part.[25]

The preceding elements of Iwade *warichi* draw attention to two processes. First is the repeated use of discussion and negotiation among shareholders. Each redistribution brought reconsideration and revision of all core elements of redistribution practice. The redistribution notebooks show only the outcomes of shareholder discussions, but it is easy to imagine the degree to which the ability to convince others and take leadership roles must have been at work. To what degree was this a function of top-down decision-making? To what degree did the outcomes benefit a limited number of shareholders or tenants? Data bearing on these issues is sufficiently complex that I take them up in separate chapters below, but the data presented so far indicate a significant role for local politicking and leadership in the management of per share joint ownership.

Second, given the supposition in the scholarly literature that joint ownership is widespread in areas of high natural hazard risk, the virtual absence of direct evidence of floods or landslides stands out. The preceding discussion has only touched briefly on the degree to which *warichi* data for Iwade village showed a sensitivity to changes in natural conditions, specifically flooding and landslides. For the most part, there was no direct evidence in the redistribution notebooks of floods or landslides as factors to which the villagers were making an adjustment. No redistribution appears to have been conducted as an emergency response to widespread flooding or a major landslide. Some changes were being accommodated as sections were redesigned during the eighteenth century, but they occurred largely in dry-field areas and were relatively subtle. The process of internal politicking will be discussed further in Chapters 7 and 8; the next chapter critically explores the important scholarly emphasis on the role of environmental considerations in joint ownership practice through analysis of multiple Echigo villages.

# 6

# *Warichi* and Natural Hazard

There has been limited mention of either floods or landslides in Iwade village thus far. If there were natural fluctuations associated with *warichi* in Iwade, they seem to have been subtle changes more than a result of significant natural calamity. The impact of environmental change, even if subtle, would have affected dry field more than paddy since restructuring of dry-field *wari* was more common than for paddy. The sole mention of floods in the redistribution notebooks occurs at the end of the reallocation of the 100-*bu wari* during the Hōreki (1756–1757) redistribution. Here we find the following note: "In the previous Year of the Ox, there was a flood, and at year-end a redistribution was conducted."

This flood seems to have affected just the one *wari*, but it is a reminder that such natural events may have been at the root of changes in *wari* structure noted in the previous chapter. Since we lack any generalized discussion of what shareholders took into consideration, we can only surmise that adjustments to *wari* reflected changes in land use by cultivators, on the one hand, and, on the other hand, changes in the total land area and quality that might result from both human actions like land reclamation and irrigation, and changes due to natural forces like floods and landslides.

It is not possible to link these shifts in Iwade *wari* directly to either floods or landslides given current sources, but it is also challenging more generally to assemble a broad evidentiary base with which to investigate directly the link between the frequency of significant flood and landslide events and redistribution. This was true even in the case of Kaga, where for much of the Tokugawa era land was reallocated only in response to a specific stimulus such as land reclamation, or, more commonly, land loss from cultivation due to flood or landslide. The linkage of *warichi* to flood and landslide risk is further complicated by the fact that the interaction of human practices—overharvesting of timber, overextension of arable land, efforts to build dikes, dams, and storage ponds—with natural forces can also create adverse "natural" events like flooding and landslides by disturbing watersheds and natural vegetation that help reduce the risk of such natural hazards. (Changes in these behaviors can also alleviate risks, of course.)

With these considerations in mind, the following discussion compares selected data from Echigo districts. The districts are small, so the climatic and other natural conditions are generally similar. In this context it is reasonable to make village-to-village comparisons.

## The Data

As the example of Iwade illustrates, in many instances there are no records that clearly state how often a village conducted a redistribution. For Iwade, one could calculate a regular interval, ten years, based on extant documents. Such instances, however, are rare since villagers tended to save only the most recent redistribution documents, not those that reveal long-term practice. Even if people made an effort to save documents for extended periods of time, to the degree that joint ownership was practiced in flood- and landslide-prone areas, repositories were at high risk of loss, compounding a general tendency for documents to disappear, be destroyed, or be abandoned over time.

In addition to cases like Iwade, where many documents that outline the principles for village redistribution practice survive and reveal the customary interval between redistributions, there are cases of oral transmission of that information. Finally, in rare instances, information has been transmitted by people who practiced redistribution in the mid- and late nineteenth century. This meant that scholars of the 1950s and 1960s still had a substantial pool of interviewees with actual *warichi* experience.

The sample data reported in local histories, surveys of *warichi* in Echigo, and similar reports come from all of the sources described above. These data form the basis for the sample of *warichi* intervals for the 113 villages examined below (sources are listed in Appendix A). This represents a fairly broad sample of villages, large enough to provide a useful foundation for examining the relationship between a region's susceptibility to flooding and landslides and the length of the intervals between redistributions. As this sample is analyzed below, readers should bear in mind that a number of villages may have followed the practice typically seen in Kaga, allocating lands only when a significant flood or landslide hit. Such villages would not have established a regular reallocation interval. Further, many local histories and depositories document the widespread use of *warichi* in Echigo but do not provide an indication of regular, periodic redistribution. Thus, the sample examined here represents only a portion of documented cases of Echigo *warichi*.

## The Hypothesis

To investigate the degree to which flood and landslide risks influenced the practice of *warichi*, I examine cases for which villagers fixed regular intervals between redistributions. If natural conditions, especially natural hazard risk, are critical in the creation and operation of per share joint ownership, the length of the interval between regularly conducted redistributions should be correlated with the propensity for flooding or landslides—shorter where such events are more frequent, longer where less common. This is not to say that natural hazards occur on a regular, predictable basis or that all stimuli to redistribute—changes in per hectare productivity, loss of land from cultivation, or adding newly reclaimed lands to the village tax rolls—were so significant that they could not be accommodated by means other than redistribution, at least over the short run. Adjustments might involve compensation for loss during the calculation of land taxes or local obligations, for example. However, where there were even small, permanent changes over time, the impact of such instances accumulated and created increasingly complex calculations. By conducting a periodic reassessment of village land and redistribution of cultivation rights, villagers consolidated the effects of the smaller changes, cleared the books, and simplified the allocation of taxes and local dues. Thus, the intervals should reflect the greater need to make cumulative adjustments from small or modest events and changes. Even in the case of incremental change rather than the loss (and possible reclaiming of) significant percentages of village land, the less stable the environment, the more frequent should be the redistributions. With that premise, I compare natural circumstances of villages in the same region to see if redistribution interval length corresponds to comparable risk of flooding and landslides and is consistent within similar geographical circumstances.

## The Provincewide Evidence

To explore the degree to which there is a relationship between redistribution interval and the natural environment, I have plotted data for villages that periodically redistributed cultivation rights on elevation maps prepared with GIS technology as outlined in Chapter 3 (base elevation data is 50-meter mesh, the highest resolution available for all of Niigata Prefecture). This approach locates villages primarily in relationship to their elevation and their relationship to drainage basins, major rivers, lowlands, lagoons, and the like. Elevation is related primarily to risk of flood damage, with those communities located low on a floodplain generally being more susceptible than communities at higher elevations.

Elevation is not a particularly good indicator of susceptibility to landslides. Landslides vary greatly in character. We typically imagine a wall of earth suddenly collapsing, but slow slumping is also a common pattern. We also often assume that allowing or encouraging the growth of vegetation on a slope will substantially reduce the potential for landslides, and in some instances that is true, especially for those involving relatively shallow earth movement. However, in other instances, the slippage is caused by water operating between soil layers that lie beneath the reach of plant root structures. There are simply too many variables to develop a down and dirty index to landslide susceptibility.

Soil type—impermeable clays, porous sands, and so on—plays some role in the potential for flooding, too. Even in conditions of similar soil, slope, and drainage area, a variety of precipitation patterns can cause flooding. For example, even in a year with average patterns of precipitation, a rapid thaw of snow can cause flooding. In other instances, slow rains over extended periods of time can saturate soils, raise the local water table, and lead to flooding. A region need not be subject to sudden, extraordinary volumes of rain to experience flooding.

Given these variables, the analysis that follows makes comparisons primarily between villages located in very close proximity to each other, that is, within a mile or so of each other. In a number of the cases discussed below, one can walk between the villages in ten to twenty minutes. By focusing on such cases, climatic differences—precipitation patterns, amounts of precipitation, and snow melt—are substantially eliminated. Difference in soil type is also minimized. In this way, environmental conditions are held substantially constant for the villages for which I discuss redistribution intervals. To the degree possible, I am comparing apples to apples, oranges to oranges. I believe this approach is useful for an assessment of the degree to which *warichi* practice is correlated with potential for change in agricultural conditions associated with floods and landslides hazards.

Data on periodic redistribution intervals have been compiled from a combination of manuscript sources, local histories (both documentary transcriptions and the survey history volumes), and scholarly books and articles for 113 villages. Using the methods described earlier for determining the location of Tokugawa era villages, I identified latitude and longitude for 110 of the villages (see Appendix A).

Plate 2 is a map showing the distribution of the villages. The sample data are by no means entirely random. Sample villages are heavily concentrated near the Shinano River, where *warichi* has been most heavily studied. Of these, many are in the vicinity of Nagaoka City. Nonetheless, the sample represents a wide variety of natural environments, from lowland plains to interior mountains, from marshy, lagoon-pocked areas to well-drained lands.

Two features of the map deserve some emphasis. First, readers familiar with this area will note that Sado Island in the Sea of Japan is missing. There is no evidence for the presence of *warichi* in this area, and Sado Island was not part of Echigo Province in the Tokugawa era. Leaving it off the map allows for a more detailed view of the rest of the region.

Second, the map is designed to focus attention on villages at relatively lower elevations. Note that the elevation legend only goes up to 67.5 meters, with the highest contour interval being 67.5 meters. Dots in the interior white portion of the map represent villages at elevations of more than 67.5 meters. Less than one-third of the sample is composed of villages at these higher levels. Most of the sample (almost three-fourths) lies at lower elevations, below 67.5 meters.

Even though most of the sample lies below 67.5 meters and only villages that practiced regular, periodic rotation of access to arable are mapped, this map makes it clear that Echigo *warichi* was not confined to lowland districts. It was present in highly varied geographic regions.

Although there are examples at the lower elevations, for example, in the low, broad Echigo Plain near the mouth of the Shinano River and modern Niigata City, classic *warichi* areas, the overwhelming majority of villages are located at higher elevations, and if they are on flood-plains, it is the narrow floodplains of smaller rivers. The common image of Echigo *warichi* as being located in the lowlands, as are the most widely studied examples, is too narrow. The geographic diversity of joint per share ownership in Echigo would be even more apparent if I had mapped all villages known to have practiced *warichi*, not just those that conducted periodic redistributions. Recall that the sample here presents only the limited number of villages for which I could identify intervals of regular redistribution. Local histories and other sources are replete with evidence of additional villages that either did not reallocate cultivation rights on a periodic basis or for which we cannot document periodic redistribution.

The diverse conditions in which *warichi* occurred in Echigo create a significant challenge in considering the impact of natural conditions on the presence, absence, and operation of land redistribution systems. Is *warichi* practiced just in the areas subject to wide-scale flooding, or must we also think about small-scale flooding? Are the frequencies of redistributions different for upland and lowland areas where *warichi* is practiced? Thinking beyond the Echigo region, if *warichi* appears in upland areas here, why does it not appear more commonly in other parts of Japan that did not practice joint ownership?

For most of the sample, data were available to document individual villages' intervals of redistribution. However, there are a small number of cases for which there are statements of standard intervals for a district, not for individual villages. In these cases (about a dozen), I have assigned villages the erstwhile district interval. This creates a somewhat artificial impression of consistency in these areas. The most prominent example is the Nadachi coastal area of modern Jōetsu City and the somewhat inland area in the same valley (Figure 6.1). The four villages in the Nadachi area have all been assigned intervals of five years; the nine upstream villages have all been assigned an interval of ten years. In still other cases, intervals for a given village were different for different categories of land (residential, paddy, dry, mountain/common land). In these instances I have used the interval for paddy in preference to the other intervals (see Appendix A for details).

For analysis the region covering the Echigo sample is divided into three general sections, very roughly corresponding to the tripartite di-

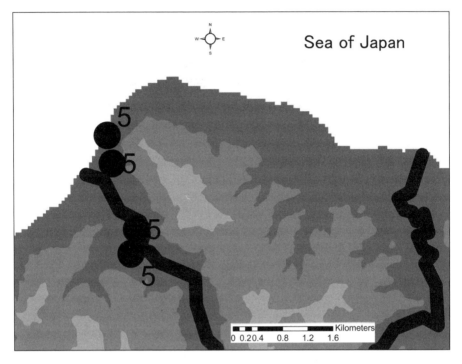

**Figure 6.1**    Nadachi Area, Jōetsu Region, Redistribution Sample

vision that Echigo denizens have long used: Upper Echigo (Jōetsu), Middle Echigo (Chūetsu), and Lower Echigo (Kaetsu), moving from south to north. Within each I present an overview map and at least one more detailed close-up. The latter are the main focus of village-to-village comparison.

Plate 3 shows a close-up detail of the Jōetsu region. To help visually compare the intervals of the villages represented, I have employed proportionally sized symbols on this map: the larger the circle, the longer the redistribution interval. The numbers beside each circle represent the actual years between redistributions. The elevation rises from sea level to about 1,500 meters, and the map employs 40-meter contour intervals. The area encompassed by the map is approximately 40 by 50 kilometers, with sample villages being confined within a somewhat smaller area. Two sets of cases for which a regional redistribution interval was applied to all the sample villages in a district are located to the left and at the vertical center of the map. The other parts of the map show considerable variation in the length of the interval between redistributions, from four to twenty years. The center of the map, where no hamlets are displayed, is an area of considerable landslide activity. It is also an area where multiple villages practiced *warichi* to deal with the effects of this hazard but did not conduct regular redistributions. Again, this serves as a reminder that the sample is an incomplete representation of total joint ownership activity.

Plate 4 focuses on the northern part of the Jōetsu region, the region in which Iwade village was located (the village of a ten-year interval just under the north-pointing arrow). The contour intervals in this map are 1-meter except for the first interval (0 to 6 meters) and the highest (white, interior areas), representing elevations of 14 meters and above. Redistribution intervals range from four to ten years (dots representing hamlet centers are uniform and not proportional to interval size). While there appears to be some consistency in intervals along the two small rivers (ten-year intervals are common), the 8-5-4-8 cluster at the center shows considerable variation in a region that shares fundamental topographical and climatic circumstances. All hamlets are at the base of relatively steep slopes, all are at a similar elevation, and all are about the same distance from the river and share a drainage basin. The 5-4-8 cluster of villages are all only about a kilometer apart. The 10-7 cluster also seems a bit anomalous, with the village situated on the river having a longer interval than the village set about a kilometer back from the river, counterintuitive if we think of rivers as the major source of flood risk.

The area just to the north of these two maps has Nagaoka City at its heart (Plate 5; two of the villages from Plate 4 are visible in the lower left-hand corner). Here 7.5-meter contour intervals are used, and the inland white areas represent elevations above 67.5 meters. Proportional symbols representing length of interval between redistributions exhibit a range from one to twenty years. Note that a number of the shorter intervals are located where we would expect them to be, right on the rivers (intervals of two, three, four, and five years in the northeast corner of the map), but some of the longest intervals appear to be along the corridor through which the Shinano River flows and is joined by other smaller rivers (e.g., the north-central part of the map, where we see twenty-year intervals).

A closer examination reinforces the impression that redistribution intervals are frequently not directly related to the risk of flood or landslide. Plate 6 is a close-up representation of the northwestern part of Plate 5. To provide detailed elevation data for the core part of this sample, all elevations of 14 meters and below are clustered in one category, and all elevations above 30 meters are grouped in one category (white). Two-meter contour intervals are used to represent elevations in between these two extremes. Why is it that in the midst of a cluster of villages redistributing every four years, one village in the same district redistributes every ten years (just below map center)? In the southwest portion of the map, where two rivers flow north to join (at Nagaoka), there is a similar disparity, with a village redistributing every twenty years, another nearby village every ten years, and smaller variations for the other nearby villages. Slightly to the north of these villages, we have four villages, two of which rotate access to land every five years and two of which rotate every ten years although the geographic conditions for all four appear very similar.

Finally, Plate 7 shows the area immediately north of Plate 6, the section of the province that includes the city of Niigata. Again, we have 7.5-meter elevation contours, and the redistribution intervals vary from one year to twenty. A noteworthy characteristic of the terrain in this section of the province is the elevated ridge that runs parallel to the coastline just north of the center part of the map. This thin elevated line extending south from Niigata is in essence a large sand dune generally no more than 25 meters high. Second, the Shinano River, which runs from the southwest corner of the map and flows north through most of the map, is today joined by the Okōzu diversionary channel (from the bottom of the map, the third stream flowing into the Sea of Japan). This channel, although contemplated in the eighteenth century, was not completed until the early twentieth century. Before that time, this was an area of low elevation but not a stream that flowed to the sea. Consequently, it is best to think of villages in these areas along the modern diversion channel as located on a plain immediately to the west of the Shinano and then ascending a ridge of over 15 meters before descending to a narrow coastal plain; they were not located along a river valley that narrows to a pass through the coastal elevated area. Within this region we will look more closely at two different areas, one near Okōzu and one to the south of Niigata.

In Plate 8, of the Okōzu area, contour intervals are of 1 meter, except for the high and low end of the scale; the highest elevation in the scale is 13 or more meters. Because there are a large number of cases in a small area and proportional symbols would overlap, I rely on a single symbol to represent hamlets and numerals to represent the period between redistributions. Even with this strategy some data are difficult to read; still, the overall picture presents several notable anomalies.

First, the number of hamlets that employ frequent redistribution but are at relatively higher elevations off the floodplain of the Shinano and other rivers (bottom, center) is quite large. This region is not prone to landslides, so that offers no explanation. Second, note the number of villages in lowland areas (about 8 to 10 meters in elevation) that are close to the Shinano (e.g., below the direction indicator) and have rather long intervals of eight or ten years, twenty years in one case. Finally, there are a number of pairs of villages about a kilometer or less apart that exhibit considerably different intervals, two versus eight years, and one versus eight years (within the southern set of villages), and five versus ten years in the northeastern pair of villages.

Plate 9 shows the region to the immediate south of Niigata City in a map with 1-meter contour intervals. The string of hamlets closely aligned along the Nishi River as it heads into Niigata are at 1- and 2-meter elevations. This is some of the lowest elevation along the Shinano, part of the area that the construction of the Okōzu diversionary channel was designed to protect. Today, protected by several diversionary channels and modern cement dikes, the area is better protected from flooding, but, during the Tokugawa era, this was a region of considerable flood risk. Indeed, the northeastern section of Niigata City, where the Agano and other rivers enter the area from the northeast, flooding was a regular phenomenon. Yet redistribution intervals here are generally more than ten years, with some as long as thirty years. A number of these intervals are the same as or considerably longer than those expected in areas less subject to flood risk.

In the preceding discussion, I have focused attention on villages that exhibit considerably different redistribution intervals but are in very close proximity to one another. Consequently, they share the same climate, often nearly identical soil characteristics, and similar slope characteristics. This is particularly true for villages located on the lower Echigo Plain near Niigata, but it also holds for a number of the examples presented in the Nagaoka and Jōetsu regions. Because of the influence of different climatic and soil variables, systematic comparison over greater distances presents a significant challenge.[1] However, there are additional analyses that can further test the impressions developed so far; they also raise an additional question.

It is possible that some places face risks not apparent in the maps examined so far, risks that other villages relatively nearby do not face. One approach to determining if this is so is to combine the redistribution interval data with geologists' modern land classifications. A second is to lay modern hazard maps over the locations of hamlets for which we have redistribution interval data. Finally, I will explore the circumstances of two adjacent villages that have very different intervals for redistribution and consider the implications of this difference when placed in the context of the data presented previously for Echigo as a whole.

## Land Classification Analysis

Plates 10 and 11 present the redistribution interval data used in the Nagaoka maps examined earlier but displayed over a land classification map. Both present a classification of the land based on soil type and other factors as well as elevation. The more subtle classification scheme is more sensitive to factors that can contribute to flooding and landslides. (Comparison with maps of elevation suggests that topographic relief is a very influential component, however.) Both maps present several pairs of cases, all within 2.4 kilometers of each other, that appear to have redistribution intervals that are counterintuitive. For example, the two villages in the center of Plate 10 share the same land classification (and the same contour interval on an elevation map) and similar surrounding environments but have redistribution intervals of ten and twenty years, respectively. In the northwest corner of Plate 11 there are two villages in very close proximity and with similar land classification that exhibit very different redistribution intervals, ten versus four years. A number of villages just below them share the same four-year interval even though they are in somewhat different land classification zones. Upland areas to the east show evidence of landslides (the dark slashes in the hillsides, Plate 11), but these are not present in the sample villages marked on this map. The primary hazard risk is one of flooding for these villages.

Although the preceding analysis focuses on just one geographically diverse district, it is apparent that geologists' classification of the terrain is not well correlated with village redistribution intervals in the pairs of hamlets that were compared. There is significant variation in redistribution interval within a classification and stability of interval across different land classifications.

## Hazard Map Analysis

In the aftermath of the great Awaji-Kansai earthquake of 1995, in order to increase public awareness of safe havens, escape routes, and natural hazard risk, as well as to improve local communities' preparations to deal with natural emergencies, hazard maps have been created in significant numbers and made available to the general public. Niigata is no exception to this trend. The focus of these maps is generally urban areas, although hazard maps have not yet been created for all major population centers. It is thus impossible to overlay the complete sample of hamlet and interval data on hazard maps.

A look at *warichi* data on the Nagaoka region hazard map is instructive. This region is not particularly subject to landslide hazard according to the local agencies that compiled hazard maps for this region. There are other potential hazards—earthquakes, for example—but they are not seen to have a relationship to joint ownership and redistribution of cultivation rights.

Flood hazard maps are in substantial measure influenced by elevation and slope within a given small area. However, these, too, incorporate other factors as well. Like land classification maps, they are more multidimensional than elevation models alone.

As with the land classification maps, Plate 12 shows a number of instances in which villages located in the same band of flood risk have very different intervals for redistribution. In the lower left section of the map, two nearby villages stand out, one with a twenty-year interval, the other with a four-year interval, but there are other cases. To be sure, there are also sections of the floodplain where intervals seem to correlate reasonably well, but there is nothing about flood risk, as defined by modern experts, that consistently correlates closely with redistribution interval.

Although this map was drawn primarily to represent flood hazard zones, various forms of slope failure are also noted on it. Among the villages displayed on the map, none are today considered at risk for major erosional events. In Plate 12, this kind of hazard is represented in the eastern, more upland sections of the map only. While modern technology can do some things to ameliorate slope failure, the sections of Nagaoka where the sample villages are located are all lowland, low-slope topography and historically not subject to landslides.

In sum, examining redistribution intervals in light of modern conceptions of flood and landslide hazard risk also fails to yield a good correlation. As with topographic and land classification maps, one finds significant counterintuitive cases.

## Shindōri and Kamegai Villages

Figure 6.2 shows a segment of a topographic map created about the time of World War I, based on the first scientific cartographic survey of Japan, completed just a few years before. It shows the two neighboring villages of Shindōri and Kamegai on the Nishi River, which runs roughly parallel to the Shinano as it makes its final journey to the Sea of Japan at Niigata. Today, this land is below sea level. In the nineteenth century, it was at most within a meter above sea level. The modern Nishi River has been channeled into a narrow, cement-lined stream beginning in the vicinity of Shindōri and ending at Kamegai, but originally there was no artificial constriction of the stream. The hamlets, both nestled on the river, are directly adjacent to each other. They share the same climatic, topographic, and soil elements. To their west is a large sand dune with a maximum elevation of just over 20 meters. Although Japan Railway bridges occasionally lose their foundation during prolonged or heavy rainfall, neither village is subject to slope failure of the dune. Both rest on alluvial soils of the same age and type. Both abut a low earth dike, a natural levee. From the standpoint of transportation, access to markets and natural resources, and other socioeconomic factors, they were identical. In the late Tokugwa era, Shindōri village was the larger of the two villages, over 1,200 *koku*, with Kamegai at more than 620 *koku*. Since boundaries for these hamlets are not available, one cannot make precise comparisons, but both villages extended beyond the map segments shown in Figure 6.2. Nonetheless, the whole area in which they are situated is one of very low relief, on the same floodplain, with only slight variation in elevation. Shindōri and Kamegai are about as close in natural, social, and economic circumstances as two hamlets can be.

Both Kamegai and Shindōri practiced *warichi*. Shindōri conducted redistributions every ten years, Kamegae every thirty years (these villages are identifiable toward the center of Plate 9 by Kamegai's thirty-year redistribution interval). What can account for such a sharp difference in interval when the socioeconomic and natural circumstances of the two villages are so similar? In discussing this pair of hamlets at a meeting of the Historical Geography Research Group of the Human Geography Society of Japan in November of 2008, one geographer who had studied *warichi* not far from this area noted that there was a strip of elevated land in Kamegai that accounted for the difference. Indeed, there is a small section of land just across the road from the

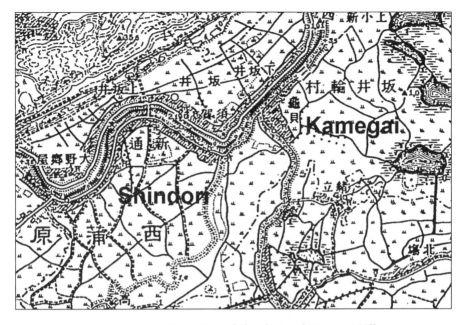

**Figure 6.2**   Late Meiji Map of Shindōri and Kamegai Villages

Japanese characters for "Kamegai" (亀貝). This elevated point is 1 to 2 meters above the surrounding fields. There may have been other small rises but on the whole no sharp difference in the two villages.

Assuming for the moment that such differences were sufficient to justify redistribution intervals that differ by a full twenty years, we then face new, and very large, questions regarding how villagers structured redistributions in other parts of Echigo. How would we account for much smaller differences in intervals in villages in other parts of Echigo that show sharper differences in topography even over very short distances? How would we account for much smaller redistribution intervals in areas that are not subject to landslides but still comprised lands with much greater topographic diversity than evidenced in either Kamegai or Shindōri?

All three of the preceding explorations suggest that the topographic maps of Echigo regions did not oversimplify the conundrum in establishing a relationship between the natural environment and *warichi*. The regional elevation maps do not mask differences that can account for the variation in redistribution interval. Close investigation of Shindōri and Kamegai and considering their practice in regional context by situating sample hamlets on land classification maps and hazard maps both reinforce the perception that there is no consistent, direct relationship between redistribution intervals and natural hazard risk or, considering evidence from the land classification map, soil endowments.

## Concluding Observations

The preceding analysis has focused on comparison of clusters of villages located very close to each other, often within a kilometer or so of each other. This approach does as much as possible to compare villages in similar natural circumstances—soil type, topography, climate, and the like—to see if villages in similar circumstances have similar redistribution intervals; the examination supports the conclusion that there is a much less direct link between land redistribution systems and villages' natural environment under joint land ownership than has been posited to date. It indicates that scholars should look more widely for explanations of the origins of proportional redistribution systems of joint land tenure than at the threat of natural calamity and variation in productivity over space or over time, the elements that have been stressed thus far. The frequency and potential severity of flooding and landslides must be seen as insufficient causes by themselves, even if we ultimately see them as necessary conditions for the proportional redistribution of arable land jointly owned. The comparison of redistribution intervals on land classification maps also suggests that natural endowments do not directly determine redistribution intervals.

Given the data at hand, there is nothing that describes the actual decision-making process that led people to develop and continue to practice per share redistributive forms of joint ownership; however, it is important to bear in mind that different individuals have different degrees of tolerance for risk.[2] Given identical stimuli, two individuals have different senses of danger and different reactions to the risks posed. An important corollary is that different communities have varying tolerances as well. Thus, in the instances above where risk of flooding, for example, appears similar, redistribution intervals can vary significantly, as with the cases of Kamegai and Shindōri villages. Conversely, in areas in which risk conditions appear quite different, redistribution intervals might be identical. The next chapters look more closely at developments, incidents, and trajectories that might contribute to the decision to practice *warichi*.

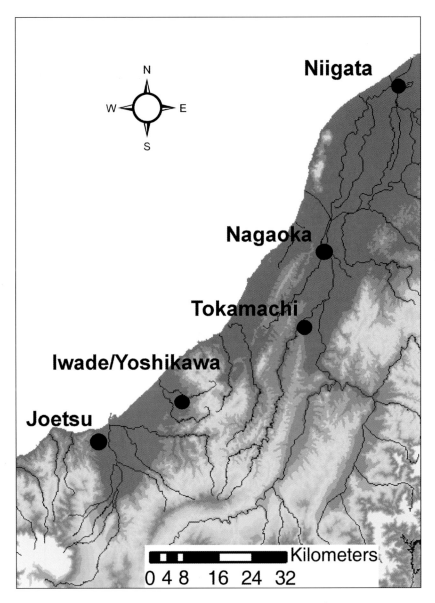

**Plate 1**  Echigo Topography and Key Research Sites

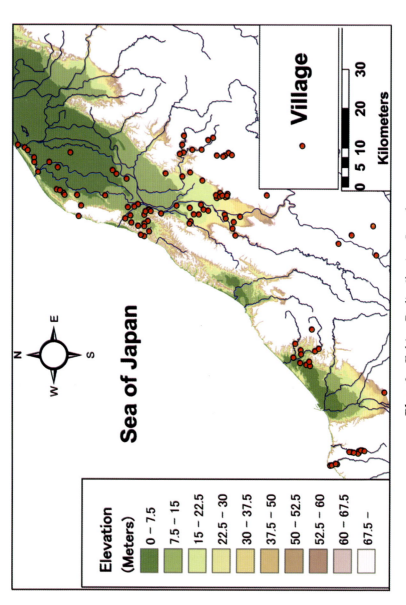

**Plate 2** Echigo Redistribution Sample

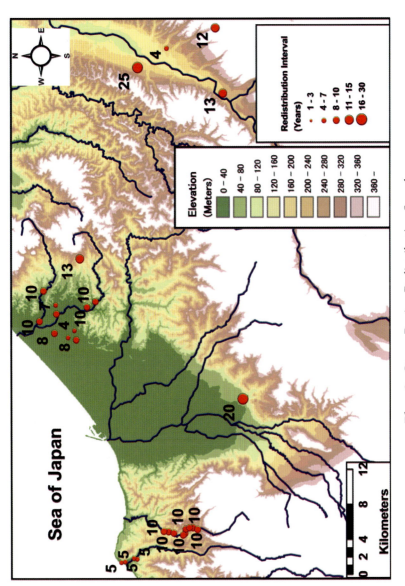

**Plate 3** Jōetsu Region Redistribution Sample

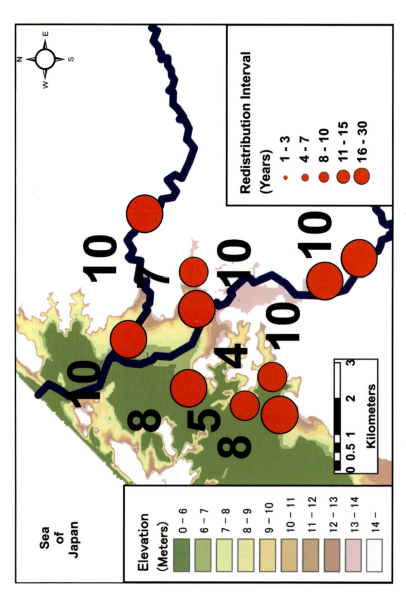

**Plate 4**  Jōetsu Region Redistribution Sample (Detail)

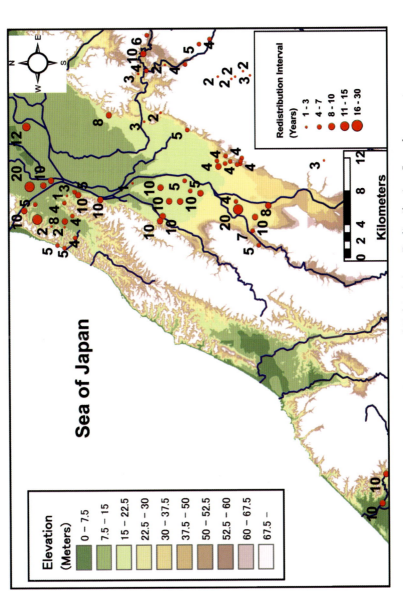

**Plate 5**  Nagaoka (Central Echigo) Area Redistribution Sample

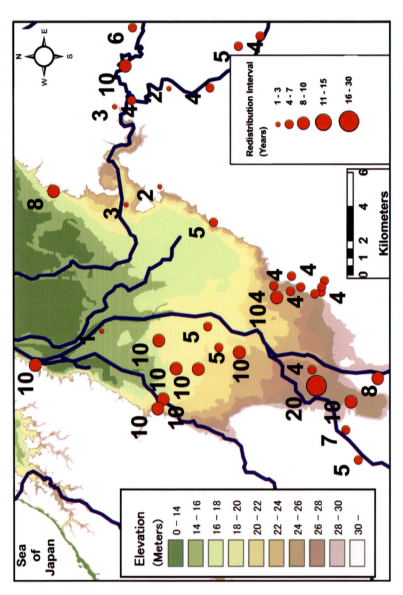

**Plate 6**  Nagaoka (Central Echigo) Area Redistribution Sample (Detail)

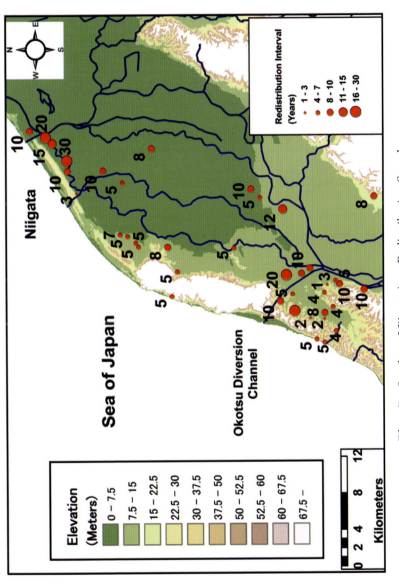

**Plate 7** Southern Niigata Area Redistribution Sample

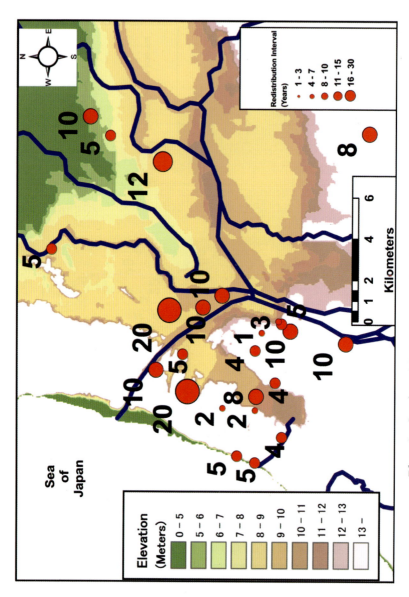

**Plate 8** Southern Niigata Area Redistribution Sample (Detail 1)

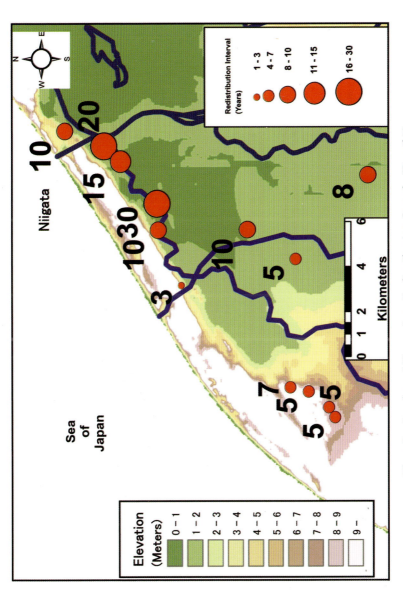

**Plate 9**  Southern Niigata Area Redistribution Sample (Detail 2)

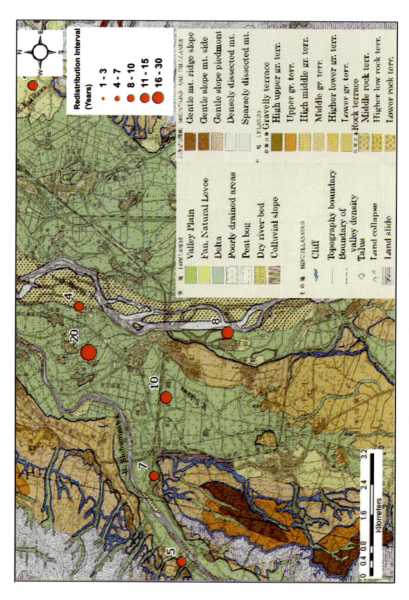

**Plate 10** Nagaoka Land Classification Map 1

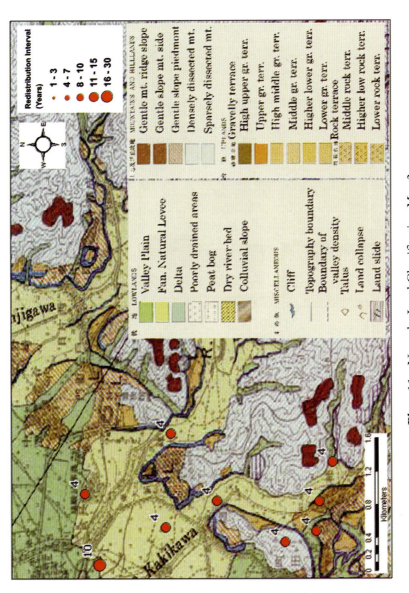

**Plate 11** Nagaoka Land Classification Map 2

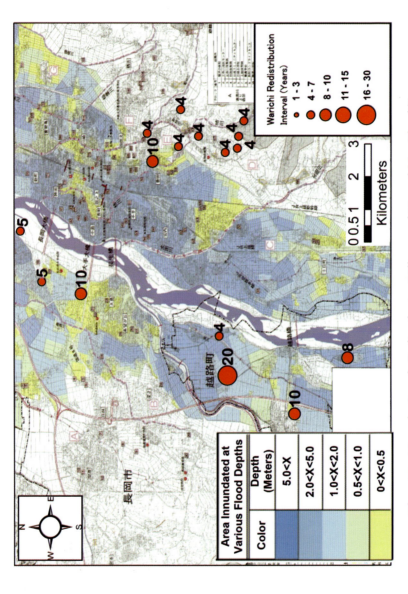

**Plate 12** Nagaoka Sample Redistribution Villages and Hazard Map Flood Zones

# 7

## Luck of the Draw?
## Outcomes and Disputes

In Chapter 5, I examined the structuring of fields and discussed the division of rights holders into groups. That discussion involved an assessment of responsiveness to the environment, a subject explored further in Chapter 6. I suggested that villagers in Iwade appeared to be responsive to changes in the quality of the soil, but data from more than one hundred villages did not show a highly regular, systematic pattern of response. There was significant variation between nearby villages in similar environmental circumstances. This suggested substantially different judgments regarding degree of risk and its import for the villages.

Shareholder judgment regarding natural hazard risk and conditions was a critical part of the *warichi* implementation, related to the process of local politicking over timing of redistributions and of land evaluation and classification. Additional questions arise about how participants approached the allocation of shares by lot and how judgment may have played a role in other parts of the process, too. How did latitude for judgment and negotiation affect fair operation of joint ownership systems?

A sense of "fairness" and "equity" is a key claim made for all forms of joint control over arable lands, and, one would expect reasonably fair operation to be a condition for village-based systems to last. For the proportional type of redistribution, the process by which lands were grouped into like categories was part of the effort to address concerns about equity, but if the mechanism for allocation of cultivator rights in each of those parcels was flawed or subject to manipulation, that foundation could be undermined. Other critical elements in evaluating parity and equity lie in the operation of the lottery and the ways in which shareholders dealt with the outcome of a redistribution.

In this chapter, I explore the impacts of redistribution outcomes, the role of aftermarkets and post-lottery manipulations that moved some cultivators to plots of land other than those they had drawn, and what disputes say about contentious issues in redistributing cultivation rights as well as domain efforts to implement *warichi*. In addition to data from Iwade village, I examine cases from several other Echigo villages and Tōdō domain's efforts to implement domainwide *warichi*. These investigations make it possible to evaluate whether or not the redistribution process was generally fair as well as offering additional insight into villager perceptions of whether joint ownership was welcome.

The investigations that follow show once again the extent to which shareholders in Iwade village reflected carefully on and negotiated each element of the redistribution process. However, some practices raise questions about the degree to which parity and equity were uniformly achieved. Still others challenge common perceptions of the desirability of different kinds of agricultural land in the eyes of cultivators. Overall, in areas where joint ownership was broadly accepted, practices show a process that was more complicated than standard descriptions indicate, not quite living up to ideals.

## Special Exemptions *(Hikikuji):* Hints of Inequities

In principle, each participant in redistribution drew lots for cultivation rights in each section of the village; however, that principle was not always followed, and some routine exemptions from redistribution were significant. For example, in Kaga domain shareholders could exclude 5 percent of their arable cultivation rights from redistribution, and shareholders overwhelmingly chose to exclude residential garden lands ( *yashikichi*). This practice protected the lands in which shareholders made the heaviest applications of fertilizer, weeded most intensively, and so on.

Like many other *warichi* regions in Echigo, in the Yoshikawa area near Iwade village (see Plate 1), all land was in principle redistributed. Nonetheless, as in Kaga, very limited amounts of land were excluded or excludable. Routinely and fully excluded lands included special lands of temples and their associated families—what we might think of as tax-exempt institutions. Such complete exclusions represented a tiny portion of villages' arable land. Beyond this practice, village headmen's land and that of district chiefs (*ōkimoiri*) might also be partially excludable as remuneration for their office work. The critical benefit in these cases was that these lands were partially tax exempt. Such an exemption provided some incentive to perform one's office duties well and to meet the demands of the overlord effectively while maintaining a smoothly functioning village community.

More commonly, and having a direct impact on all shareholding members of villages in the Iwade area, some communities began their distribution of arable land by implementing a rule focused on so-called residential lands, the garden and seed plots. Such rules specified the exclusion of fixed amounts of such lands—some villages excluded 35 *bu*— from the redistribution for each participating household. (I have not been able to identify such a clear standard for Iwade village.) Where this practice was followed, the residences of dependent cultivator (*nago*) families as well as independent farmers were included.[1]

To understand this practice, readers must first bear in mind that in premodern Japanese villages houses typically were tightly clustered. There were exceptional regions such as the "scattered residential villages" (*sankyoson*) of Tonami County (*kōri*) in Etchu Province, modern-day Toyama Prefecture, but as a rule, the clustering of houses limited the opportunities to expand one's garden plots. Except on the outer reaches of the residential cluster, one would have simply bumped into one's neighbors' lands.

Nonetheless, not everyone had the same amount of garden and seedbed lands, and some method was required to implement parity in residential lands excluded from redistribution. To this end, the area around the house was measured, excluding paths, up to the main road. This measurement defined the area of a household's excludable garden and seedbed land that was classified as "residential." This area was then subtracted from the allotment for exclusion. If, owing to some special condition or circumstance—for example, initial unequal lot sizes that were grandfathered into the rules when the practice was begun—the amount of land contiguous to the residence exceeded the allotment and it was not possible to partition it off to another holder, then that amount of land was subtracted from the registrant's access to superior paddy in another section of the village, one removed from the residential center of each village. This alone would be a sharp disincentive to expand one's garden lands even if there were such an opportunity.[2]

Conversely, if there was not enough land contiguous to the house itself, compensating acreage was to be provided from another superior dry-field section of the village. In these circumstances, the shareholders not only received the amount of land they were short, they received a bonus allotment of 20 percent of the area of the detached holdings. This premium provided compensation to the recipient for the inconvenience of having to travel a greater distance to work land that was separated from the residence.

Since the amount of land available for expansion near the residential sections of a village was very limited, all these adjustments represented a marginal portion of a cultivator's holdings. As with the process by which villagers determined field grades and characteristics when they classified land for redistribution, shareholders appear to have proceeded based on discussion and creation of a consensus; the issues were not so problematic that their resolution was recorded for future reference. With these limited exceptions, all land was in principle subject to redistribution.

In Iwade, customs superficially fit this general tendency, but in fact they are much more puzzling, presenting further exceptions to these rules and hinting at special treatment. It is not that I could not identify any fixed area or amount of residential land that was routinely excluded from redistribution in Iwade village; the phenomenon of *hikikuji* (withholding lands) I am about to explore falls entirely outside such a pattern. While the term literally means "pulling or drawing a lot," in this instance it means withholding from redistribution a plot of land that one had managed up to the current redistribution. The person who held this right could keep land cultivated heretofore, and in the current redistribution no one else had the opportunity to receive the right to cultivate it. This right was limited to one redistribution only. Holding the right in one iteration of a redistribution did not convey the same right in successive redistributions. According to Funahashi's brief analysis, the Satō family used these procedures to retain control over specific fields it desired, in particular, the Waseda and Himetsuruta paddy *wari*.[3] This point was not central to Funahashi's analysis, and thus a variety of issues remained unaddressed. Was the option of withholding shares from redistribution available only to the Satō family, the largest landholder in the village, or was it open to other families? Was it possible to apply *hikikuji* to dry fields as well as paddy? Were these rights repeatedly exercised in the same sections of the village?

Table 7.1 reveals several salient characteristics of exemptions from redistribution. Rows record the year of distribution, the document number, the total number of exemptions from redistribution *(hikikuji)*, the number of paddy exemptions, the number of dry-field exemptions, the number of different shareholders exercising exemption rights, and the number of sections of land *(wari)* in which the exemptions were exercised. Since some shareholders held multiple shares, there could be more shares than shareholders exercising exemption rights. Conversely, since there were some partial shares, it would theoretically be possible for the number of people exercising exemption rights to outnumber the number of shares with which such rights were associated.

The data in this table indicate something more complex than just one family's effort to retain control of the same field(s) year after year. Further, this process of excluding shares from redistribution differs considerably from a broadly generalized right of all participants to exclude, say, 5 percent of land from each redistribution as was common in Kaga. This right was restricted to a portion of shareholders.

First, note that the number of exemptions differs from the total number of shares in the village. A shareholder exercising *hikikuji* did so on just a small fraction of the sections *(wari)* of the village within which he drew lots. Remember that each holder of a full share would, in principle, draw lots for lands in each of twenty-four to twenty-eight different sections of the village (see Chapter 5), so shareholders exercising exemptions on one or two draws out of more than two dozen were exercising this right on a small proportion of all the lands to which they had cultivation rights.

Second, the number of exemptions *(hikikuji)* differed considerably from redistribution to redistribution. Typically, two to three people exercised these rights on a half dozen to a dozen subsections of fields. In general, one person withheld from redistribution the land for only one or two subsections per *wari*. The key exceptions are 1809 (Appendix B, Table 2) and 1818 (*wari* 3 and 4, Appendix B, Tables 3 and 4). Among those lands over which shareholders chose to exercise *hikikuji* (Table 7.6 below shows this most clearly), the Gotanda paddy *wari* (both 1 and 2), the dry-field Asabatake number 1 *wari*, and Shinden *wari* were most frequently chosen. Nonetheless, the broad scattering of lands over which participants chose to exercise *hikikuji* indicates that each did so based on his own style of managing his entire agricultural enterprise—styles that appear to have changed over time—and no individual family created a consistent safe haven in one or two sections of the village.

**Table 7.1** Cases of Exempted Lands *(Hikikuji)*, 1710–1848 plus Undated

| Year | Document Number | Number of Hikikuji Exemptions | Number of Paddy Hikikuji | Number of Dry-field Hikikuji | Number of Shareholders Exercising Exemption Rights | Number of Wari in Which Hikikuji Rights Were Exercised |
|---|---|---|---|---|---|---|
| 1710 | 8006 | 2 | ? | ? | 3 people: Hachibei, Zenbei, one unspecified share (that of the document author?) | 1: 1 dry-field (?) *wari* (50-*bu wari*) |
| 1746 | 8013 | 0 | 0 | 0 | | |
| 1710 | 8372 | 3 | 0 | 3 | 1: Yūshichi | 1: 1 (Asabatake 2) |
| 1756–1757 | 8009 | 14 | 5 | 9 | 1: see Tables 7.2, 7.3 | 7: 3 paddy, 4 dry |
| 1765 | 2554 | 0 | 0 | 0 | | |
| 1781 | 8014 | 10 | 5 | 5 | 1: see Appendix B, Table 1 | 6: 3 paddy, 3 dry |
| 1800 | 8373-1 | 3 | 2 | 1 | 2: Denbei-Gen'ya plus 1 name unclear | 3: 2paddy, 1 dry: 1 in Uranaka 1, 1 in Uranaka 2, 1 in Asabatake 1 |
| 1808 | 8373-2 | 17 | 7 | 10 | 2: Takahara 3, Zenbei 1 | 4: paddy (Nokori 50-*bu* 2) |
| 1809 | 8374 | 56 | 21 | 35 | 14 (+ 5 unclear): see Table 7.4; Appendix B, Table 2 | 15: 9 paddy, 6 dry (+3) |
| 1818 | 8012 | 29 | 19 | 10 | 3: Hachibei, Zenbei, Denbei; see Table 7.5; Appendix B, Table 3 | 12: 8 paddy, 4 dry |

| 1818 | 8375 | 44 | 15 | 29 | 9: see Table 7.5; Appendix B, Table 4 | 14: 5 paddy, 9 dry |
|---|---|---|---|---|---|---|
| 1840a | 8377 | 0 | 0 | 0 | | |
| 1840b | 8378 | 0 | 0 | 0 | | |
| 1842 | 8379 | 11 | 11 | 0 | 8 (7):* see Appendix B, Table 5 | 5: 5 paddy |
| 1848a | 8380 | 0 | 0 | 0 | | |
| 1848b | 8381 | 0 | 0 | 0 | | |
| unclear | 6954 | 4 | 4 | 0 | Gotanda 1 (3), Kitarō; Waseda (1), Yūshichi** | 2: Gotanda 1, Waseda |
| unclear | 8007 | 0 | 0 | 0 | | |
| unclear | 8008 | 1 | Unclear | dry? | Yūeimon | 1; 100-*bu wari* dry? |
| unclear | 8015 | 0 | 0 | 0 | | |
| unclear | 8016 | 0 | 0 | 0 | | |
| unclear | 8017 | 8 | 8 | 0 | 1 (perhaps cultivator, not shareholder), see Appendix B, Table 6 | 3: 3 paddy (of which one is the author's seedbed) |
| unclear | 8264 | 12 | 12 | 0 | 5: see Appendix B, Table 7 | 5: 5 paddy |

*Note:* Bold rows indicate data extracted from this table for further analysis in Table 7.6.

* One suspects that Sōzaeimon, Yūshichi's man, was not drawing a lot on his own behalf, and that would mean that seven shares exercised *hikitokuji*.

** Kitarō exercised three exemptions in the Gotanda 1 *wari*; Yūshichi exercised one in the Waseda *wari*.

*Source:* All documents from Satō-ke monjo.

Reflecting on these two points, it becomes clear that any rule that governed the withholding of land from redistribution was narrow in focus, permitting only a small fraction of village land to be withheld. Such rules may also have specified conditions for withholding lands that could not be met by any individual consistently from iteration to iteration. For example, the right to withhold land might have been contingent on the cultivator having made some improvement to the land. It is also possible that there was no rule per se and that the opportunity to withhold land from redistribution varied based on a case-by-case reconsideration, the outcome a product of discussion and consensus building during each redistribution. In either event, the outcomes show that no shareholder used exemptions to withhold consistently the same land from cultivation and that the total amount of land withheld represented a limited portion of total village arable.

The proportion of all land that comprised *hikikuji* can be calculated only roughly. It varies from a 1756–1757 low of less than 2 percent to a high in the 1818 *kujikaechō* of 14 percent.[4] High figures were exceptional, and the proportion of land subject to *hikikuji* fluctuated, never exhibiting a sustained upward trend or a permanent focus on particular sections of the village; it was both very temporary and unstable. Nonetheless, the general trend was for the number of *hikikuji* to grow (or, if there was a rule that defined contingencies under which land could be withheld, those contingent circumstances became more common) over the late eighteenth and early nineteenth centuries. Since the dates of some documents are unclear, one must be cautious in reaching this conclusion, but certainly by the nineteenth century, the amount of land on which shareholders exercised exemptions from redistribution was greater than the 5 percent permitted all Kaga domain shareholders, and the rights to withhold were concentrated in a few hands, not universally shared. The notion that *hikikuji* was not granted to all participants departs significantly from standard understandings of how *warichi* operated.

In light of the emphasis on equity in descriptions of redistributional practices, it is worth stressing that Table 7.1 indicates that rights to withhold land from redistribution were limited to a small portion of shareholders. In years when the number of cases of withholding was small, only one to three shareholders had such rights. In years with many such cases, fourteen different households—almost all shareholders—had at least some access to this privilege. In 1756–1757, the sixteen shares were organized into nine lot-drawing groups (*kujigumi;* two families held more than two shares each), eleven families participated, but only six exercised rights to withhold land. In 1809, nineteen families participated, of which fourteen held such rights.[5] In 1818, the same number of families participated, and thirteen withheld land (there are two documents for this redistribution with slight differences between them; for this analysis I have not tried to resolve the differences but have simply noted the data presented in the original notebooks).[6] The average number of withholdings *(hikikuji)* per person changed considerably over time: in 1756–1757, 2.33; in 1809 (Bunka), 3.5; and in 1818 (Bunsei), 9.6 or 1.5 depending on which document one looks at.

While it appears that during each redistribution shareholders determined anew who had rights to withhold lands and the sections of *wari* in which they could exercise those rights, evidence available does not permit any assessment of the reasons for most decisions to allocate *hikikuji* to individuals. The exemptions are simply noted. Any additional explanation concerns only their location.

It is difficult to make a direct comparison of the number of *kenmae* shares held and the number of *hikikuji*, but the evidence suggests there was no connection. Tables 7.2 to 7.5 show that in a number of instances the combined number of shares and *hikikuji* exemptions do not agree when compared shareholder to shareholder (or by totals, where that figure can be calculated). Access to the right to exempt land from redistribution was not determined as a proportion of shares of cultivation rights held.

Furthermore, although a number of documents are incomplete or present data only for a partial redistribution, where it is possible to confidently make a comparison in Table 7.1, *hikikuji* exemptions tend to

**Table 7.2**   *Hikikuji* Exemptions vs. *Kenmae* Shares, 1756–1757

| VILLAGER | NO. OF *HIKIKUJI* | NO. OF SHARES |
|---|---|---|
| Sukeueimon | 1 | 1.5 |
| Yūeimon | 6 | 6 |
| Tazaeimon | 2 | 2 |
| Yūshichi | 1 | 2.5 |
| Yūeimon, Yokichi, Riueimon | 2 | 1 |
| Yokichi | 1 | 0.5 |

*Source:* Satō-ke monjo, document 8009.

**Table 7.3**   *Wari* Distribution of *Hikikuji* Exemptions, 1757

| | Paddy | | | Dry Field | | | | |
|---|---|---|---|---|---|---|---|---|
| Villager | Ura-naka I *W'RI* | Go-tanda I *W'RI* | Sawada 70-*BU* *W'RI* | Asa-batake 30 *BU* 1 *W'RI* | Asa-batake 30 *BU* 2 *W'RI* | Shin-den-batake 60-*BU* *W'RI* | 100 *BU* 1 *W'RI* | Total |
| Sukeueimon | | | 1 | | | | | 1 |
| Yūeimon | 1 | 2 | 1 | 1 | 1 | | | 6 |
| Tazaeimon | | | | 2 | | | | 2 |
| Yūshichi | | | | 1 | | | | 1 |
| Yūeimon, Yokichi, Riueimon | | | | | | 2 | | 2 |
| Yokichi | | | | | | | 1 | 1 |
| **Total** | 1 | 2 | 2 | 4 | 1 | 2 | 1 | 13 |

*Source:* Satō-ke monjo, document 8009.

appear more frequently on dry field than on paddy or in approximately equal proportion. This impression is surprisingly counterintuitive since paddy is generally presumed to have a high value relative to dry fields, and one would think that paddy would therefore be the object of withholding. However, given the fluctuation in the number of dry-field and paddy *wari* sections from redistribution to redistribution previously noted, this impression deserves closer investigation.

Data from the boldface lines in Table 7.1 are extracted in Table 7.6 to facilitate comparison between *hikikuji* use on dry field and paddy. The calculations in Table 7.6 add further weight to the impression that exemptions were disproportionately exercised on dry fields. To the data from Table 7.1 I have added calculations of the number of *wari* sections as presented in Chapter 5. The notebooks for the redistributions in this table are clearly complete, and they contain sufficient detail to provide an accurate comparison. Study of these cases provides the best way to demonstrate that dry fields were favored when shareholders exercised their *hikikuji* exemptions. (The 1710 document has a number of unnamed and unidentified *wari* sections that interfere with analyzing the use of exemptions, and the documents from the 1830s and 1840s are not complete, so all are excluded.)

**Table 7.4**  *Hikikuji* Exemptions vs. *Kenmae* Shares, 1809

| Villager | No. of Exemptions | No. of Shares |
|---|---|---|
| Zenbei | 9 | 2 |
| Zenbei, San'ueimon | 1 | 1 |
| Sakuzeimon | 1 | 1 |
| Kitarō | 16 (18) | 4 |
| Shōzaeimon, Yohachi, Kitarō | 1 | 1 |
| Sakuzeimon, Fujizaeimon, Kitarō | 1 | 1 |
| Yasuzaeimon | 3 | 1 |
| Sahachi | 1 | 1 |
| Senzō, ? | 1 | not recorded |
| Yohachi, Shōzaeimon | 2 | ? |
| Yōzaeimon, Shōkichi | 1 | 1 |
| Denbei | 2 | 1 |
| Takahara | 1 | ? |
| Yoshiueimon, Jūbei | 1 | 1 |
| Unclear | 7 (12) | ? |

*Source:* Satō-ke monjo, document 8374.

**Table 7.5**  *Hikikuji* Exemptions vs. *Kenmae* Shares, 1818, *Gechō*

| Villager | No. of Exemptions | No. of Shares |
|---|---|---|
| Hachibei | 23 | 4.5 |
| Zenbei | 12 | 2 |
| Denbei, Yoshiueimon | 1 | 1 |
| Eitarō | 1 | 1 |
| Sahachi | 1 | 1 |
| Hyōeimon, Naozaeimon | 1 | 1 |
| Shōzaeimon, Yahachi | 2 | 1 |
| Kishichi, Yūeimon | 1 | 1 |

*Source:* Satō-ke monjo, document 8375.

Table 7.6 Distribution of Exemption Use in Paddy and Dry Field, 1756–1818

| Year | Data Source (Document No.) | Stable Paddy W/RI | Unstable Paddy W/RI | Z'NBU Paddy | Total No. Paddy W/RI | Non-Z'NBU Dry-Field W/RI | Z'NBU Dry | Total Dry-Field W/RI | Total No. W/RI | Hikikuji Exemptions, Dry Field | Hikikuji Exemptions, Paddy | Total Hikikuji Exemptions | % Dry-Field Hikikuji Exemptions | % Dry-Field W/RI Sections |
|---|---|---|---|---|---|---|---|---|---|---|---|---|---|---|
| 1756–1757 | 8009 | 12 | 1 | | 13 | 6 | | 6 | 19 | 5 | 9 | 14 | 35.71 | 31.58 |
| 1781 | 8014 | 12 | 4 | | 16 | 7 | 1 | 8 | 24 | 5 | 5 | 10 | 50.00 | 33.33 |
| 1800 | 8373 | 12 | 3 | 1 | 16 | 9 | 1 | 10 | 26 | 2 | 1 | 3 | 66.67 | 38.46 |
| 1809 | 8374 | 12 | 3 | 2 | 17 | 14 | 1 | 15 | 32 | 21 | 35 | 56 | 37.50 | 46.88 |
| 1818 | 8012 | 12 | 1 | 2 | 15 | 9 | 1 | 10 | 25 | 19 | 10 | 29 | 65.52 | 40.00 |
| 1818 | 8375 | 12 | 2 | 3 | 17 | 8 | 2 | 10 | 27 | 15 | 29 | 44 | 34.09 | 37.04 |
| Total | | | | | | | | | | 67 | 89 | 156 | | 42.95 |
| % of total hikikuji | | | | | | | | | | 42.95 | 57.05 | | | |

Source: Table 7.1.

Note that for every document in Table 7.6 the number of paddy *wari* outnumber the number of dry-field *wari*, often by substantial margins, up to 220 percent more (1756–1757, thirteen paddy, six dry). The percentage of all exemptions exercised on dry field for all six redistributions is approximately 43 percent even though the proportion of all *wari* sections comprising dry field is below that level for all but one document (1809). With the exception of 1809, the proportion of all exemptions exercised on dry field exceeds (three redistributions) or substantially equals (two redistributions) the proportion of all *wari* sections comprising dry fields.

As seen with the careful attention to the configuration of dry-field areas of the village when establishing land classification divisions for redistribution (Chapter 5), the emphasis on dry field suggests its importance in the lives of farmers. For one, where specific reasons for *hikikuji* are identified below, a number of them point to the need to produce vegetables, a key dietary component. Furthermore, if standard understandings of Edo period villager diets are accurate and rice made up a very small part of consumption, the emphasis on dry fields as the site for exercising exemptions suggests a concern for daily food supplies, as opposed to the production of rice, a good primarily used for land-tax payments. Finally, there was no major commercial crop to occupy a prominent place in farm management; thus the importance of foodstuffs and therefore dry-field crops.

Equally interesting is the evidence that, when shareholders exercised the right to withhold land, they chose different places at different times. People did not repeatedly target the same area. This raises the possibility that shareholder motivations for protecting some of their lands from redistribution changed over time. This conclusion is hinted at in the far-right column in Table 7.1, but it is fully manifested in Table 7.7 and the tables in Appendix B. For example, in Table 7.7, an exemption was exercised in the Gotanda 1 *wari* section of paddy in 1757, but no one exercised an exemption in that *wari* in the next two redistributions. While the two Gotanda and two Uranaka paddy *wari* are consistently sites at which exemptions were exercised, the other paddy *wari* (more than half the total paddy *wari* where exemption rights were exercised) were used by exemption rights holders only sporadically. Except for 1781, no exemptions were exercised in the Kaminishi 2 or Kaminishi 30-*bu wari* sections of village dry field. In only one dry-field *wari* were exemptions exercised in at least four redistributions, the Shinden (dry field) 60-*bu wari*. Only five of the fifteen dry-field *wari* were the

**Table 7.7**  Distribution of *Wari* in Which *Hikikuji* Exemptions Were Exercised, 1757–1840, plus Undated

| | PADDY | | | | | | | | | |
|---|---|---|---|---|---|---|---|---|---|---|
| | 60 BU (WASEDA) | URANAKA 1 | URANAKA 2 | GOTANDA 1 | GOTANDA 2 | SHIMONISHI 100 BU | SHIMONISHI 25 BU | ZANBU 50 BU 2 | HIMETSU-RUTA 1 | SAWADA 70 BU |
| 1757 | | 1 | | 2 | | | | | | 2 |
| 1781 | | | 1 | | 3 | 1 | | | | |
| 1808 | | | | | | | | 4 | | |
| 1809 | 1 | 1 | 2 | 6 | 5 | 1 | 1 | 1 | 1 | 3 |
| 1818 *kujikae* | 1 | 2 | 2 | 4 | 4 | 1 | 1 | 4 | | |
| 1818 *gechō* | 1 | 2 | | 4 | 4 | 1 | | 4 | | |
| 1840 | 1 | 2 | 2 | 4 | 2 | | | | | |
| Ms. 8017 | | 2 | | 3 | 3 | | | | | |
| Ms. 8264 | 1 | 1 | | | | | | | | |

Dry

| | ASA-BATAKE 30 BU 1 | ASA-BATAKE 30 BU 2 | SHINDEN 60 BU* | 100 BU | KAMINI-SHI 30 BU | KAMINI-SHI 2 | NOKORI-HATAKE 30 BU | TAKE-HARA-HATAKE 50 BU 1** | TAKE-HARA-HATAKE 50 BU 2** | TAKE-HARA-HATAKE 50 BU 3** | TAKE-HARA-HATAKE 100 BU 1 | TAKE-HARA-HATAKE 100 BU 2 | TAKE-HARA-HATAKE 50 BU 1 | NAGA-TORO-HATAKE 45 BU | KAWA-HARA-NOKORI-HATAKE 40 BU |
|---|---|---|---|---|---|---|---|---|---|---|---|---|---|---|---|
| 1757 | 4 | 1 | 2 | 1 | | | | | | | | | | | |
| 1781 | | | | | 1 | 2 | 2 | | | | | | | | |
| 1809 | 10 | 5 | 3 | | | | 2 | (2) | (2) | (3) | | 4 | 4 | | |
| 1818 *kujikae* | | | 4 | | | | 2 | | | | 3 | 1 | 2 | | |
| 1818 *gechō* | 9 | 5 | 4 | | | | | | | | 3 | 2 | 2 | 1 | 1 |

*Sometimes called Hatake 60 *bu.*

**Names of exemption holders in these wari were crossed out; nonetheless, I have treated them as actually exercised exemptions.

*Sources*: Satō-ke monjo, documents 8009, 8012, 8014, 8017, 8264, 8373-2, 8394, 8375, 8377, 8378.

object for exemptions in three redistributions; nine of the fifteen *wari* where exemptions were exercised were the site for exemptions two or fewer times.

The preceding data speak to the use of *hikikuji* in the village as a whole; what of its use by individual farmers? The tables in Appendix B display how individual shareholders used their right to exempt land from redistribution. They indicate that individual shareholders exercised their exemptions across a broad portfolio of *wari* sections rather than concentrating them in one or two, even when an individual held a significant number of *hikikuji* exemption rights. For example, in 1809, Kitarō exercised ten exemptions over three different paddy and perhaps eight more over four different dry-field *wari* sections. In the same redistribution, Zenbei exercised two exemptions on two different paddy *wari* and seven over four different dry-field *wari* sections. Both men also participated in exercising additional exemptions for shares held jointly with others, further diversifying the portfolio of *wari* on which they exercised their exemptions.

The pattern of exemption usage cannot be parsimoniously and completely explained, but we do have explanations for a number of cases. These cases provide insight into shareholder thinking even if they cannot be generalized to all usages. Do these partial explanations suggest inequity, or are they grounded in fair principles?

In some cases, exemptions were part of the process that assured participants access to a fixed portion of the residential lands, those used as garden plots and seedbeds, and clearly fit within the boundaries of fair practice. For example, at least a part of the *hikikuji* exemption exercised in dry-field areas in 1818 was directly linked to Hachibei's residential lands. The 1818 *gechō* notes, at the end of recording the allocation of sections in the Takehara 50-*bu wari*: "In the preceding thirty-two places, the eight numbered 25 to 32 are Hachibei's exempted lands [*hikikuji*]. Of these, 300 *bu* are part of his residential lands; another 100 *bu* presently are part of Kaneshichi's residential lands." In the same document, in describing the allocation within the Hatake 60-*bu wari*, we find the notation, "Of the preceding thirty-two places, 720 *bu* are within Hachibei's residential lands."

The implication is that within lands classified as residential, there were dry fields that were removed from redistribution—as all residential lands appear to have been by the nineteenth century. Recall that the link between dry field and residential lands was clearly established in the early notebooks that treated paddy, dry field, and residential lands.

Because of the link to an individual's residential land, this form of *hiki-kuji* was compensatory and not a special privilege. This sort of *hikikuji*, then, does not appear to represent any violation of principles of equity or fairness in redistributional practice.

The two notebooks for Bunsei further suggest that some *hikikuji* served a very different function; they were associated with land reclamation and as such were part of an incentive open to all. In the *gechō* entry for Takehara 100-*bu* number 1 *wari*, section number 15, the scribe notes, "In the 40 *bu* contiguous to Sekiyama, there is land that is now paddy"; in the Takehara 50-*bu* number 3 *wari*, section number 21, he notes, "Presently, seven fields are paddy"; and, finally, in the Asabatake 30-*bu wari*, section 1, he wrote, "Drawn as residential land, at present this land is paddy." Each of these cases was treated as *hikikuji*.[7]

These cases are all associated with a form of land reclamation called *hatakenaoshi*. Rather than create completely new arable, in this form of reclamation cultivators made investments in water supply, diking, and the like to convert dry field to paddy. Such conversions were treated separately for tax purposes. The cases just introduced all appear to be individual efforts at reclamation, not joint village efforts. This distinction is important. Had these efforts been the result of a collaborative village effort, there would be no particular need to withhold the newly improved lands from redistribution. All could have been distributed to the collaborators by, for example, treating them as a separate *wari* with sixteen subsections and reallocating it as they would any other *wari* (as in Iwade village's Shinden *wari*).

Incentives to create or improve arable land operated at the individual or small-group level as well as the villagewide level. Incentives were personal—the opportunity to increase and benefit from the amount of land one farmed—but there were also tax provisions that encouraged both individual and collaborative efforts by small groups or whole villages. Special domain tax-exempt treatment of reclaimed or improved land was universal in Tokugawa Japan. Domains lowered taxes on newly improved lands—providing an incentive to undertake the labor investments essential to further develop arable land—in effect treating the short-term loss as an investment that would yield increased land-tax revenues in the long run. This treatment was extended to lands developed individually or collaboratively.

In *warichi* regions, whether in Echigo or in other parts of Japan, the rules of the redistribution typically recognized the investments that an individual or small group made by exempting such lands from redistribution. These exemptions were usually temporary. Individual efforts at land improvement should be considered distinct from instances when a whole *wari* was composed of reclaimed land, as, for example, Iwade village's Shinden *wari* (literally "reclamation *wari*") which was entirely reclaimed by the collaborative efforts of all the shareholders in the village. In any case, exemption from redistribution, like the tax exemptions for reclamation, was part of the overall incentive structure to encourage the expansion and improvement of arable lands.

While both reclamation and lands treated as part of a shareholder's allotment of residential lands (garden plots and seedbeds) explain a limited number of *hikikuji* exemptions, many other cases lack discernible explanation. We must entertain the possibility that some sort of inequity underlay this practice. We have a partial explanation for *hikikuji* exemptions but not one that is fully satisfactory.

The number of people holding exemption rights in Iwade was rather large, ranging from about one-third to two-thirds of shareholders. These special rights were not at all stable, and locations varied considerably over time. Whatever principle operated to determine rights to exempt land from redistribution, its use in practice was transient enough to suggest a response to specific, temporary conditions that might apply to any shareholder and applied widely enough to suggest that any shareholder, not just the largest shareholders, had the chance to benefit under appropriate circumstances,. The redistribution notebooks demonstrate that no shareholder was on a trajectory of accumulating permanent rights to exempt lands from redistribution. Nor did anyone consistently exercise them in the same sections of land, redistribution after redistribution.

## Aftermarkets

Funahashi's analysis of the role of *warichi* in the Satō family's management of their farmlands reveals another interesting, postredistribution wrinkle. He notes that there were clear instances in which the Satō exchanged the lot they drew with another family in order to continue to cultivate a section of the village that they had cultivated up to that time. For example, Funahashi notes that in the 1757 register the Satō family traded section number 12 in the number 2 *wari* (*ni no wari*), which they had drawn, to Yūshichi and Rieimon in exchange for section number 15 (which these men had drawn), a section that the Satō had cultivated up to this time.[8]

In other words, once the drawing was complete, participants were able to engage in aftermarket transactions. No rule prohibited such private transactions. There is no indication in this particular case that money changed hands, so other factors—the preeminent position of the Satō in the village, family ties, ease of access for the families involved, mutual benefit or something similar—provided the incentive or social compulsion that underlay these exchanges. Such adjustments created conditions for the shareholders that they felt were more favorable than the original outcome of the draw.

## Incentives for Special Crops

From the standpoint of *warichi*'s potential to retard economic growth or to accommodate commercial crops, how long-maturing crops with potential commercial value were treated holds particular interest. If such investments were discouraged, then the system acted to discourage the growth of commercialized agriculture and the expansion of a cash economy. For example, crops such as the lacquer tree grew for a number of years before reaching maturity and harvest. Certain varieties could produce sap that would be refined and used to make lacquerware. Other varieties produced wood for construction, kindling, charcoal, and other uses. Although primarily for local consumption, even these trees had commercial potential. Both uses required an investment horizon longer than one or two agricultural cycles.

Ozawa village (Yoshikawa area), not far from Iwade village, planted only one such crop, *urushi*, a Japanese variety of sumac or lacquer tree. The variety produced here was that used for construction and other purposes, not for the production of sap used in lacquerware. Although its commercial value might be less than the sap-producing trees, treatment of land planted in lacquer trees suggests the degree to which *warichi* could be adapted to accommodate commercial crops. The settlement of a dispute in 1773 reveals the following basic principles for dealing with such crops.[9] (1) Only trees above a certain size were included in the redistribution. This assured the original holder that his initial investment of labor would not come to nothing because of a redistribution (when trees were young and exempted, how long one could continue to use the land on which these trees were planted is unclear). (2) Before land was turned over to the new cultivator, all trees were to be harvested by the original cultivator. (3) If there was mutual benefit to not cutting trees, they could remain for an additional year. This judgment was left to the individuals involved. (4) However, if there was no agreement, "when the redistribution period is exceeded, the land must be cleared within two months, and if there is a dispute, the trees become the [new] landholder's [unless there is prior mutual agreement]."

While the first condition might appear to punish the cultivator of the lacquer trees, that is not the case. If we consider the conditions as a whole, it is clear that trees above a certain size were considered harvestable and of sufficient size to enable a return on cultivator investment if not to maximize investment return. Trees above the designated size were to be harvested and used or sold by the cultivator. The third condition suggests that there might be benefit to letting the trees mature further, but in that instance, some accommodation needed to be reached with the new cultivator, presumably a sharing of the benefits the harvest would bring, since the new cultivator would lose the use of this share of his land. That the size limit was considered sufficient to cover the costs of the investment is suggested by the fourth condition, which placed the new manager of the land in a stronger position in negotiations with the prior cultivator—if no agreement could be reached, the new cultivator got 100 percent of the benefit of the trees.

This arrangement demonstrates institutional support for some long-term agricultural investments; however, cultivator initiatives were restricted. As already noted, residential garden lands (*yashikichi*) often included not only the land on which houses were built, but also certain dry-field lands. Some enterprising cultivators converted residential lands into paddy entirely on their own initiative. The dispute settlement just cited specifically instructed that these lands be reconverted to dry field. At issue was not simply keeping a particular category of land sacrosanct or cultivator opposition to economic advancement. This order responded to the critical need to continue to supply water to established paddy in the village. Where water supply was uncertain, as was the case in parts of the Yoshikawa area, an upstream resident who took water for his newly converted land typically compromised downstream residents' ration of irrigation water.

Water supply issues were involved in privately converting dry field and mountain land to paddy.[10] Such restrictions, which were frequently in place throughout Japan, were intimately tied to the nature of paddy agriculture and were not the specific product of *warichi* practices. That some conversions of this sort were accommodated is confirmed by the presence of paddy in some Iwade village *warichi* sections that were otherwise dry field.[11] These fields remained in the dry-field allotment be-cause they were generally considered to have the same value as the dry fields they replaced; that is to say, for tax purposes they were considered inferior or at best average paddy.

## Disputes and Contention

The evidence on incentives discussed so far comes from document-ed dispute resolutions. Discussion of disputes reminds us that *wari-chi* practitioners had to exert themselves to make these arrangements work. Given the propensity to idealize arrangements like *warichi* (as with the Russian *mir* and similar organizations), concrete reminders of the efforts needed to maintain these social arrangements are espe-cially useful. Once a dispute moved outside the confines of a village and into the realm of higher administrative authorities, records of dispute resolution reveal how domain authorities responded to village custom. The discussion below indicates that rather than use such disputes as a lever with which to expand their ken, officials generally, but not always, sought to identify and respect past village practice and precedent.

Even considered just at the village level, a comprehensive system of joint land tenures would have engendered more conflict within villages than we can ever document. The process of measuring and grading land could take several weeks, and although farmers in each village knew the characteristics of the arable land in great detail, there must have been quibbling over what land to include in which category, how big each plot was, and whether the division of each section of the village into comparably productive *wari* was appropriate. Such debates were typical-ly resolved internally, without invoking the authority of external powers.

In spite of the potential for conflict, I have yet to discover a single documented case of purely intravillage conflict over *warichi* procedures or outcomes. The dispute documents I have examined focus primarily on the question of who was able to participate. Village officials were even known to lie and deny they had a history of using *warichi* despite earlier records that, if revealed to officials, would have documented the practice.

Participants, especially those with money, had the option of ap-pealing an intravillage settlement to district or other officials, but they appear to have done so infrequently. Why do we lack records of in-travillage disputes? Using a process of elimination, we can first rule out communal harmony. When nonresident shareholders brought lawsuits, they often found allies among the resident shareholders, strongly suggestive of intravillage factionalism.[12] When someone else was prepared to take the lead, fellow travelers were waiting to capitalize on the opportunity.

All dispute cases I have found to date—a small handful of cases—involve a primary litigant who was either a nonresident landlord or someone else viewed as an outsider by villagers.[13] Why wait for an outside leader? Several social and political factors played an important role. First, although old landed wealth had lost much of its grip on smaller shareholders since the early seventeenth century, hierarchical relationships within villages were still strong. Even when parvenus challenged old wealth for a share of political power, that did not necessarily expand the base of political rights substantially, nor did it mean that smaller shareholders gained a base for autonomous political action. Many were still beholden to their wealthier counterparts in some way. In a society as stratified as Japan's, intravillage politicking could involve potentially significant threats to one's livelihood, and these kinds of pressures should be remembered as we consider the paucity of evidence on intravillage disputes.

Wealth may have played a role in a very different way, too. At each stage of such an investigation, villagers had to bear the expenses of the investigating officials' visits. Furthermore, in many parts of Echigo, at least after the early eighteenth century, it was common for shareholders to provide gifts to district officials (*ōkimoiri* and others), county-based officials of the domain *(daikan)*, and so forth. These gifts, in addition to basic meals and lodging, had to be provided to any official who visited a village, and the gifts were quite rigidly scaled to the rank of the official. When villagers requested an inspection to lower taxes or to investigate a dispute, these gifts were mandatory.[14]

Gifts represented a significant expenditure, and the way of dividing contributions among villagers tended to penalize those who brought a suit or requested an inspection. The largest share was borne by the plaintiff, with the balance borne by the other shareholders. As is true today, justice was most readily available to those who could pay for it; early modern Echigo shareholders had to be able to afford the costs of carrying a dispute outside the village. Not only were these costs substantial, even the division of that financial burden sometimes led to disputes, further complicating an ongoing case![15]

In sum, multiple factors encouraged protesters to coalesce around an outside leader. Such leaders typically were wealthy and socially and financially independent of the locally powerful. They could provide necessary funding as well as an alternative policy position vis-à-vis the dominant parties in the village.

Dispute records that involve nonresident shareholders reveal some additional significant characteristics of Echigo *warichi*: the need for village officials' concurrence to redistribute. In one instance, it is clear that a single, intermediate-level village official *(kumigashira)* was able to stop a request for a new redistribution. The village headman *(shōya)* deferred to his objections. In another instance, the failure of another *kumigashira* to sign off on a request hindered but did not stop one Tomizaeimon from pressing on with a direct appeal to a district official *(ōkimoiri)*. Many would have lacked the chutzpah to carry on with this enterprise, but Tomizaeimon was a man of considerable resolve. He ultimately made a direct appeal to the temples and shrines magistrate of another domain, taking his petition directly to Edo. For this he was placed under a light form of arrest and was ultimately fined—very light penalties by contrast with the common image, purveyed in historical literature, of imposition of the death penalty for direct appeals.[16] Once the fine was paid, Tomizaemon again took up his cause with local officials. Such protests were unusual, however. Yet even in this instance we see the rule in operation: to press for a redistribution (either as an emergency response to a major calamity or when one had not been implemented for some time), the unanimous consent of village officials was essential.

Finally, the dispute resolutions I have examined suggest that, where a tradition existed, outside authority was prone to support its continuance, even when that involved supporting the claims of nonresident shareholders against a united village officialdom. In none of the cases discussed above did authorities do anything other than uphold existing practices. In this regard, the decision of the Meiji government's Office of Civil Administration's ruling in the Iwade village dispute (Chapter 5) was right in line with Tokugawa practice.

The preceding examples concern village disputes brought to district and regional authorities for resolution; however, domain efforts to influence or implement *warichi* could create difficulties. These were likely to be manifested as the domain attempted to implement joint ownership realmwide or to promote major changes rather than in relationship to a single village's redistribution. In Kaga domain's early-nineteenth-century efforts to regularize the intervals of redistribution, resistance was passive, not active; however, in other cases, the onset of domain efforts resulted in direct and open opposition.

Tōdō domain, in what is today modern Mie Prefecture, presents a very different case from that of passive resistance in Kaga, one of active opposition to domain efforts to implement *warichi*. In this case, villages thought it antithetical to their interests. Before the domain efforts, a number of villages, especially in the upland areas of the domain, practiced village-based proportional *warichi*. In effect, domain authorities were attempting to extend its use in a domainwide, standardized fashion. Some details of Tōdō's proposals are unclear, but this policy was implemented in tandem with fiscal reforms and land sale policies that many villagers resisted. The issues associated with sale of land rights involved the magistrate *(kōri bugyō)* Ibaragi Rihei's effort to cut through increasingly entangled layers of multiple mortgage obligations on a plot to identify who was responsible for payment of land taxes and to limit the tendency for wealthy villagers to accumulate in their own hands more and more cultivation rights to the best village farmlands. To restrict the accumulation of such rights, Ibaragi planned domainwide implementation of joint landholding and redistribution, and he planned to do so in a way that actually redistributed landed wealth, taking away cultivation rights from larger holders to allocate them to others. Interestingly, this policy also provoked the opposition of some domain retainers and intendants *(daikan)*. Initial efforts at the end of 1796 called for implementation in thirty-eight especially impoverished villages, with redistributions to take place decennially to maximize the potential for small holders to survive and not lose all of the best farmlands to the wealthy. Before the end of the year, strident opposition (an *ikki*) broke out; the first demand of the village protesters was for an end to *warichi*. As a result of the protests, efforts to measure and assess fields ground to a halt, and shortly, the entire effort was abandoned. Midlevel villagers were the most strident opponents of *warichi*, apparently concerned to protect their own possibilities for accumulating good farmlands, but small holders also opposed the proposed land reform. The issues involved here extend beyond the joint form of land tenure and include an emphasis on redistribution of wealth; nonetheless, the redistribution system drew sufficient concern to lead to effective opposition to domain policy.[17]

## Conclusion

Villagers tampered with the luck of the draw in a number of ways. Some manipulations would not be considered violations of equitable functioning of *warichi:* provision for equitable allocations when some individuals had too much or too little residential land, preservation of incentives that promoted investment in agriculture (whether in the form of extending cultivation, improving fields, or planting crops requiring more than a single season to reap return on investment), or post-redistribution private trades. A large number of *hikikuji* exemptions cannot be explained so benignly, however. The discussion in this chapter leads to the impression that there was a measure of inequity in the operation of *warichi* in Iwade.

Quite apart from the routine functioning of *warichi,* this chapter has examined several instances in which *warichi* operations became the focus of legal disputes. Given the role of "outsiders" in legal cases, the perspective they present is limited. We can see that relations with "outsiders" were strained and that people like Tomizaemon could be very adamant. In these cases, the issues in dispute were not always directly related to the implementation of a specific redistribution but involved the question of whether nonresident shareholders would be included. These outsiders had sufficient means to be able to finance a suit, at least working in concert with other villagers. Nonetheless, in combination with the discussion of the apparent degeneration of *warichi* in Iwade in Chapter 5, these cases reinforce the impression that coordinating and continuing the practice of joint control over arable lands was a challenging task. The case of opposition to a new, domainwide policy associated with *warichi* as in Tōdō domain, however, represented a more generalized complaint that united villages against domains rather than dividing villages internally. Tōdō's experience indicates that *warichi* was not always a venue for collaboration between village and domain interests.

Both the disputes and the variant practices display an ongoing tension between participants' sense of private interest and the functioning of a system of joint land ownership that sometimes contravened that interest. In some cases, like Tōdō's, that conflict was between village communities and the domain, but more typically the contention was between those owning cultivation rights within a village. In some instances, the issue was not whether to implement or continue the system but the details of how a system on which people agreed in principle should function in a given instance, the exercise of exempting lands from redistribution *(hikikuji)*, for example. While shareholders recognized a need to cooperate though joint ownership and while they may have been altruistically motivated in some measure, they did not thereby sacrifice self-interest. This is evident in the examples of variance from commonly described joint ownership principles observed in Iwade village—aftermarket exchanges and the like.

The evidence in this chapter hints at the degree to which a combination of leadership, social pressure, and good social skills was essential to the successful functioning of joint landownership. Additional evidence presented in the next chapter reinforces this impression.

## 8

# Adaptability, Survivability, and Persistent Influences

Joint ownership arrangements served multiple functions, and systems could be fine-tuned in sophisticated and sometimes complex ways. Earlier chapters have examined *warichi* in one village in considerable detail and explored whether there was a relationship between the flood and landslide risks posed by the natural environment, on the one hand, and the sensitivity of redistribution to differing degrees of risk, on the other. Even within per share systems of joint ownership, the most common form, community behavior across the single region of Echigo displayed varied responses to natural hazard risks. This chapter takes up three additional subjects: the adaptability of joint ownership systems over time, the place of tenants under joint ownership, and the related issue of its persistence into the twentieth century. All three discussions indicate both a commitment to joint ownership and thoughtful consideration of how it could be adapted to serve new ends or function in new social, economic, and political environments.

Because of limited time-series data, much literature on *warichi* does not provide extended-term examination of change over time. We are left with a rather static image of joint landownership arrangements. However, as historians are well aware, human institutions are seldom static. They adapt to changes in the social, economic, and political environment. When I explored the adaptability of *warichi* over time in one village, Iwade, that discussion took place in the context of a single

171

political-economic order. Although there was a nationwide movement toward greater rural participation in an increasingly well integrated national economy, a change in which Echigo participated, there was not a broad, rapid shift in the economic regime in most of early modern Japan.

A fundamental transition accompanied the Meiji Restoration and Japan's fuller integration into the global order of late-nineteenth-century imperialism. This circumstance fostered radical changes in the economy that continued in postwar Japan: growth of town-based factory manufacturing as well as restructuring of land taxation under a modern, centralized state. How did *warichi* fare under these new conditions? Was it flexible enough to adapt to the new economic and political order? To what degree could it be adapted to new functions? What insights does its evolution at this time yield, not only about the transition to a new market and political order, but about joint ownership's use before the Meiji Restoration?

The adaptability of joint ownership is partly tied to one additional subject we have yet to discuss: the place of tenants in these regimes. If all cultivators were shareholders, we could reasonably stop at this point; however, as is universally recognized, many early modern and modern cultivators were either purely tenant farmers or owner-tenants who cultivated some lands that they owned outright as well as lands that they rented from others. These two groups represented the vast majority of the agricultural population in late-sixteenth-century to early-twentieth-century Japan. Even if the per share redistribution worked to provide a diversified portfolio of lands to shareholders, what evidence is there that it worked for tenants? Are there indications that this method of controlling access to arable lands worked for them? Does exploration of the participation of tenants in *warichi* suggest any further insights into why it was employed, why large landlords tolerated or actively supported it and how the system changed over time?

Serendipitously, the best materials for exploring both the role of tenants under *warichi* and its potential for flexibility and adaptability to special or new circumstances comes from the transition into Japan's modern era, primarily the nineteenth century and beyond. This documentation reveals a new element of fairness and equity not yet discussed—the way in which rents were fixed. This and other characteristics tell a good deal about why joint ownership survived well into the twentieth century.

## Adaptation to Changing Circumstances

I have largely discussed *warichi* as though it functioned in the same way over time in a given location, acting as though its functions and outcomes in Echigo did not evolve. Most cases cited in secondary literature convey the same impression, but such was not the case in practice. Here I introduce examples to demonstrate the potential for villagers to adapt *warichi* over time, transforming its function even in the context of a single community.

Data from the area of modern Tokamachi City, upstream from Nagaoka along the Shinano River (see Plate 1), show that, early on, *warichi* in this area was directly associated with the ravages of the Shinano and the vagaries of weather. A 1790 contract from Asabatake village notes, "Because there are a number of swampy areas amongst the fields, each instance of flooding brings inundations of rock and mud, landslides, or breaches in the river banks. At the upper reaches of streams, there is damage from a lack of water during years of low rainfall. On these occasions...all households without distinction participate in *warichi* and reallocate cultivation rights to fields by lot, helping all in the village."[1] Villagers saw redistribution here as a means of distributing natural hazard risks among all shareholders in the village, and this document suggests that it was implemented as needed, based on natural events and conditions rather than periodically. Such motivations were common in villages along the Shinano.

Two villages examined here, Minami and Kita Abusaka, are on opposite banks of the Shinano River as it flows through Tokamachi. Kawaji, a third case, lies about a kilometer from the river on a tributary, more upland. Unlike downstream areas, the floodplain is narrow, and hills rise steeply from the riverbed. The flooding of the river historically caused dramatic changes in the area of arable land cultivated near the river. Given the varied topography within even a single village, some land in a village was exposed to this risk, and the rest was not.

To begin, I look at a case where villagers decided to limit *warichi* use to a particular part of the village, discontinuing villagewide implementation. Kawaji first implemented *warichi* right after a 1682 domainwide survey that touched many Echigo villages. Dissatisfied with the official survey results (formal surveys of large areas of the domain were often hurried affairs, subject to a variety of errors), shareholders conducted their own to properly allocate the land tax burden among themselves—a common pattern in the annals of *warichi* practice (recall the explanations for its origins in Echigo).[2] Under these circumstances, villagers' desires to get the internal division of taxes right is entirely understandable even though their work could not force the domain to recalculate the village assessed value *(kokudaka).*

Kawaji went a step further and also implemented the first documented iteration of what became a regular village practice of per share joint ownership. The village's assessed value was determined by the domain to be 391 *koku,* and this the villagers divided into shares that allocated lands to some forty families. Each received access to land in proportion to the size of their presurvey shareholding rights. As in Iwade village, this process resulted in each shareholder managing rights to an identically structured portfolio of arable lands that reflected the benefits of soil fertility, access to water, and so on, as well as all of the natural risks, whether in the form of inadequate water, low fertility, or flood and landslide risk. Unlike Iwade village practices, Kawaji shareholders from the start determined that residential garden and seedbed lands would be excluded entirely. Residents set regular four-year redistribution intervals. Over time, the village experienced repeated flooding as rushing waters broke through dikes. Shareholders laboriously cleared inundated and damaged lands of stones, mud, and debris, and then duly reincorporated them into the redistribution process.

This custom continued until 1767, when eight shareholders undertook to convert unfarmed lands into productive arable. They petitioned village leaders to exclude these lands from redistribution. The implication underlying their request is that they were carrying a special burden not shared by all in the village and that, by including the lands once again in the redistribution, they would effectively provide benefits to free riders who had not participated in the reclamation.[3] A vigorous debate followed. As noted previously, exemption of newly improved lands from redistribution had precedent in the region. Temporary exclusion of reclaimed lands was one way to provide incentives for individuals and small groups to improve the land. If the motivation here was to secure temporary exemption of the land from redistribution, this request would not be noteworthy. However, in this instance, others in the village argued that the lands that were "improved" had actually once been part of the village's arable and that this was not newly reclaimed land, but a restoration of arable that had been damaged by flooding. This, they argued, meant that the lands should not be excluded from redistribution.

The document on which the preceding summary is based (see

note 3) leaves many questions unanswered. However, it appears that the lands damaged (and restored by these families) were in a very small section of the village, a small portion of a whole *wari* subdivision to which these individuals had drawn lots. Only under such circumstances would opposition to a request for exemption from redistribution make much sense.

Although we cannot be certain that it bears a direct relationship to the 1767 dispute, a document two decades later suggests that the discussion over where to implement *warichi* within the village continued. In 1786, the village shareholders declared that the village would no longer conduct redistribution for all arable lands. Redistribution was henceforth to be conducted only on the older, established lands *(honden)*. Newly reclaimed lands were henceforth to be excluded in perpetuity.[4]

Whatever the specific outcome of the settlement two decades earlier, it appears that the incentives to bring land (back) under the plow were not sufficient. By this declaration, the village sought to remedy that situation. Ultimately it determined that inducements targeted at individuals or small groups were the appropriate approach to expanding the village's arable land. With these inducements, village leaders acted as though it was easier to get an individual or small group to restore, expand, or improve land than it was to get all shareholders to cooperate in such efforts. They rejected claims of those who insisted that, as in the past, even privately reclaimed lands be brought within the purview of the village's *warichi* practices.

What might have been responsible for this division of opinion, this change of approach? Frequent loss of the same land to floods on a repeated basis offers one possible explanation. Some cultivators may simply have decided that this land was too much at risk to merit continued collaborative support. Given Kawaji's location between the river and a hillside that rises rapidly to more than 200 meters, a second possibility is that the newly reclaimed lands now coming under the plow were all uplands, not subject to the significant natural hazard risk and again not brought into production by joint effort of all shareholders.

Regardless of rationale, Kawaji shareholders and their village leaders determined that a two-tiered approach to redistribution of cultivation rights was now in order. While the long-established fields would remain under periodic redistribution, henceforth newly improved lands would not be reallocated. Villagers saw no fundamental contradiction in employing two systems of ownership rights simultaneously and as complements to each other.

The Kawaji village discussions tell of the flexibility that villagers found in the ways they implemented *warichi*. While emotions must have run high in these discussions, change was clearly possible. Other cases from this same region indicate that other villages exhibited similar creativity and flexibility.

For example, one Naizōsuke opened fields in 1813 in Minami and Kita Abusaka: in the latter hamlet, authorities surveyed six fields of inferior paddy totaling more than 5.2 *chō* with an assessed value of more than 57 *koku*; and in Kita Abusaka, they surveyed five fields of inferior paddy totaling more than 5.27 *chō* and valued at almost 60 *koku*.[5] This development took place on an island in the Shinano River formed by the deposition of erosional soils, an area that was the subject of a dispute between the villages of Tokamachi and Minami Abusaka. The area was ultimately divided between the two communities in an 8:2 ratio, as a result of which the larger area became known as the "Eight Parts Island" (Yatsubun Shima) or just "Eight Parts" (Yatsubunkata).[6]

Naizōsuke put up the capital to develop the area beginning in 1808 and was apparently quite successful—for a while. In 1826, a flood wiped out almost all of the reclamation project, and he had to begin pretty much from scratch. This project continued to be unstable, and it was not easy to keep cultivators engaged in cleaning fields up, rebuilding them after floods, and returning the fields to cultivation. A contract signed in 1855 indicates that Naizōsuke sought to alleviate this situation by devoting a quarter of his farmland to use by people who either had no fields to cultivate or had very little land under their own management.[7]

Here Naizōsuke employed the promise of allocating cultivation rights to unlanded or small-holder farmers in an effort to attract people with a strong incentive to continue to farm this property, people who had only limited alternative options. But this incentive was also linked to efforts to protect their stake in these fields, since all of these lands were subject to reallocation through periodic *warichi* redistribution. In other words, to the degree that there were different cultivation risks associated with different sections of the island, they would be shared equally by all stakeholders. Such an arrangement limited the losses of any given family if floods swept away part of the lands and simultaneously ensured that everyone had a vested interest in rebuilding after a disaster. In this case, the use of *warichi* served two of the functions of joint landholding outlined in earlier chapters—allocation of benefits from joint reclamation enterprises and natural hazard risk amelioration.

In addition to providing some of the risk-diversifying and invest-ment-sharing functions typically associated with *warichi*, Naizōsuke's initiative shows that per share joint ownership of land was fully com-patible with the self-seeking effort of individual entrepreneurs to ex-pand their own wealth. Naizōsuke could put up the capital needed to open or reopen these fields, but he lacked a motivated labor force. By sharing access to the land and implementing per share joint owner-ship and redistribution, he provided incentives to attract labor and a portfolio structure of land rights that was as diversified as this location permitted, a structure that maximized the possibility for individual cul-tivators to survive the periodic inundation of the area.

His own pecuniary considerations aside, Naizōsuke's decision to allocate part of these lands to landless and small shareholders started a practice that continued long after his direct management of these fields ended. Over the next several decades this land and the use of joint own-ership and redistribution came to serve something of a social welfare function. While the land was never available to cultivate for free, it was always allocated first to the less fortunate in the village.[8]

The transition to ultimate village takeover from Naizōsuke and a greater welfare function began later in 1855, in the eleventh lunar month, when both Minami and Kita Abusaka finally ended their dis-pute regarding the segment of the island over which they had fought for years: they agreed to administer the disputed lands jointly.[9] Reclaimed land identified in the 1855 document was a small area, valued at just over a *koku*, but the principle of per share proportional allocation of the revenues from it to the two villages was clearly established in their agreement. Income was now at least partially directed to the villages, not to Naizōsuke. While the tie between these lands and Naizōsuke's family remained in 1855, it was finally broken completely in 1868, the year of the Meiji Restoration.

For the first years of the new era, the structure of land rights in Japan remained as it had in the Tokugawa era. Yet before its first dec-ade was out, the government imposed a new system of taxation and private right of possession based on the British model. The new policy laid down the principle that lands were either privately held, commons *(iriai)*, or belonged to the state (where there was no written documen-tation of private or common land rights).

Abusaka villagers discussed how to treat the lands in Yatsubun Shima under the new tax laws, and, in effect, they agreed to super-ficially comply with them but to subvert their intent in practice. All shareholders on the Yatsubun Shima lands were issued certificates of landownership, and all documents submitted to supravillage authorities did nothing to suggest than that the new Meiji tax laws had not been implemented as ordered. Official documents available to county, prefectural, and national authorities indicated that these lands were now held, fee simple, by individuals and not jointly owned.

However, the shareholders of Yatsubun Shima also acted to protect their preexisting customary system of joint ownership. In late 1875, they signed an agreement among themselves that accomplished that objective. Although this document refers to these lands as "commons" (*iriai-chi*), a phrase that was not used in documents to describe this land before, signatories did not alter the earlier practices for managing cultivation rights. They did make some decisions about how to mediate between the new tax structure and their traditional practices. They determined that taxes on the land would be divided equally. To make this accord with village tax records, all shareholders were listed in village records and certificates of ownership issued for equal proportions of the hamlet land. Further, to counteract the privatizing thrust of the new Meiji land tax regime, they agreed that no individuals would act on their own to open new lands to cultivation, to plant crops like trees on the land, or in other ways to remove fields from agricultural use. Any decision to open or improve land or to remove it from cultivation remained the corporate prerogative of the shareholders. Through this compact, shareholders effectively continued joint ownership of the land.[10]

In Yatsubun Shima, compliance with the Meiji Land Tax Reforms was simply a fiction. Meiji government officials had all the documentation they required to indicate that their new system of landownership and taxation had effectively been implemented, while Abusaka villagers had an agreement in hand that clearly showed that nothing of the sort had happened. All their traditional methods of dealing with the land remained undisturbed!

Joint ownership under erstwhile private property continued to evolve here (and in hundreds of other villages; see below). An 1881 agreement among shareholders on cultivation of the Yatsubun Shima area indicated that all cultivators were now volunteers and that those signing the agreement retained cultivation rights for four years, beginning in 1882. The village would pay the land taxes due and collect a uniform fixed rent for the period from each cultivator. The village would not be responsible for any land lost to floods or other natural acts. In other words, the village government was fully cooperating in maintaining the fiction of private ownership. While participants dropped out and others were added over time, the principle of giving access to the landless and small owners continued, and *warichi* survived in this part of modern Tokamachi City. This arrangement continued through the prewar years and well into the postwar era. Such joint corporate enterprises, while increasingly rare, continued here and there into the 1970s.

## The Fate of *Warichi* after the Meiji Land Tax Reforms

In principle, the Meiji land reforms of the 1870s eliminated joint landownership and other complex land rights.[11] In July 1873, the new government issued the ordinances compelling land taxes to be based on the cash market value of land rather than its presumed yield *(kokudaka),* paid in cash, and assessed on individuals rather than on the village as a whole. Multiple and joint community ownership of arable land were to cease.

Orders to implement the new land tax system duly arrived in the Echigo region (entirely encompassed by Niigata Prefecture by 1876). Despite the description of Yatsubun Shima above, the new regime had a clear impact on *warichi* practice. After the land reforms of the 1870s, the practice declined significantly, especially over the long run.

Nonetheless, as Yatsubun Shima's experience indicates, the practices of village control of land and cooperative payment of land taxes continued for some time in the Echigo region. Examples can be found in other parts of Japan, too, and, at least for the first decades of Meiji, such examples were rather common, certainly more common than what is conveyed by standard images of effective, comprehensive, and immediate implementation of new land tax laws.

Individual elements of village control of arable land survived in diverse parts of Niigata well into the twentieth century in at least several hundred villages. Even where villages abandoned redistribution among landowners, many of its practices continued to govern tenant-landlord relations. The Meiji government made no quick, rigorous, and forceful attempt to eradicate these local customs. As in Yatsubun Shima, local officials selectively enforced the land tax reforms to accommodate local practices, thereby helping to keep opposition to a minimum and permitting remnants from Niigata's "feudal" past to shape landholding and tenant practices.[12]

Although certificates of private landownership were issued, redistribution and joint responsibility for national land-tax payments continued in a minimum of 239 villages through at least 1887.[13] Yorobigata village south of Niigata continued joint payment at least until 1899.[14] As late as 1928, one authority could count a dozen contemporary administrative districts (comprising about thirty-three Tokugawa villages) in which redistribution of ostensibly private land still occurred.[15] In exceptional cases, such redistribution even continued well after the land reforms of the postwar American occupation. Thus some villagers continued to treat land taxes as a joint corporate, rather than individual, responsibility well into the twentieth century. At least in areas of frequent flooding, the risks of natural disasters and taxes continued to be shared by residents.[16] For other farmland identified as "commonly held" *(kyōyūchi)*, land certificates were often not handed out at all, and redistribution practices continued unabated into the first two decades of the twentieth century.[17] In these cases, land taxes on individual plots were added together, and the villagers computed their proportional share based on traditional *warichi* practices.

Even in those villages in which private landownership was established, customs associated with the old system of land management conditioned land sales and tenant relations. For example, it was common for deed transfers and rental agreements through the early twentieth century to describe amounts of land transferred in accord with old standards of land value employed under *warichi* (*kyū mura nami warichi X kenmae*, or "so many *kenmae* shares of the old village *warichi*")—*kenmae*.[18] Local public assessments, too, often continued to be based on these traditional units of land measure.[19]

## Landlords and Tenants

Nothing in joint per share ownership as practiced in Echigo forbade or inhibited rental of land. In these areas, as in other parts of Japan, land rentals could be used to alter the size of a farm enterprise, depending on labor available as families moved through different cycles, increasing the area they cultivated when there were more robust adult laborers available and decreasing the amounts of land cultivated when labor was in short supply. Similarly, if branch families were created, they would typically be provided with limited amounts of land, land that needed to be supplemented by renting land from the main family.[20] And for those with more land than they could cultivate with their own family labor, renting land out was a logical option, especially in an era when day wage labor was limited.

Consequently, shareholders in Echigo villages often included families, like the Satō in Iwade village, large landholders who rented out a considerable portion of their cultivation rights to others. As residents and shareholders of the village, they participated in the planning and implementation of redistributions. Thus, the Satō drew lots in each of the implementations documented in Chapter 5.

Yet the Satō were also large enough landholders that they possessed rights to manage many lands across multiple villages. They were hardly unusual in this regard. Throughout early modern Echigo, a good number of large landlord families owned lands in villages other than that of their residence. Such families typically lived relatively close to the villages in which they held lands during the Tokugawa era. In many instances, they could readily walk to villages where they had tenants.[21] Tokugawa period absentee landlords often retained roots in or near the districts in which they held cultivation rights.

Based on their nearby residence, one might presume that Edo era absentee landlords participated in the redistribution, but this appears not to have been the case, or at least not universally so. For example, a landlord who controlled lands rented out to residents of Kawaji village apparently held some lands that were treated as outside of the redistribution.[22] And in Asabatake village in 1790, an absentee landlord specifically petitioned to be allowed to participate in the redistribution.[23] The outcome of the petition is not known, but along with the evidence from Kawaji, we must conclude that absentee landlords did not automatically participate in a village's redistribution practices.

If there were landlords, then by definition there were tenants. While *warichi* notebooks are the only direct documentary descriptions of actual reallocation outcomes, they stop with allocations among shareholders. We are left with the question of how access to fields was allocated to tenants. Tenants within Iwade who held no shares at all and shareholders who also rented land cultivation rights from others participated in a second draw. Although we lack notebooks or similar documents that describe a redistribution of cultivation rights to tenants via lottery, the scattered pre-Meiji evidence for tenant participation in combination with the widespread use of the practice by tenants well into the twentieth century indicates broad tenant participation and strong attachment to *warichi*.

Why should tenants be asked or demand to participate in redistribution of cultivation rights? If shareholders, acting as farm managers, were concerned about sharing all the benefits and risks of village lands, then it is hardly a big leap to suggest that tenants, also acting as managers of a farm enterprise based in whole or in part on rented lands, would have the same interest. For the poorer tenants, whose farming businesses were more economically marginal, the interest in maximizing access to good or excellent lands in the village as well as lower-quality land might well have even been stronger than for their wealthier counterparts. The village of Kawaji, discussed above, provides a pre-Meiji example. The dispute settlement of 1786 specifically notes that tenants were to be included in the *warichi* process.[24] The same document notes that rents were to be set villagewide.

A question raised in Chapter 5 remains: why did landlords, especially those as dominant as the Satō of Iwade village, go along with such arrangements? To the degree that they did go along and not seek to prevent tenant redistribution of cultivation access, it was part of their effort to retain adequate labor to cultivate their lands. In this regard, their interest paralleled that of daimyo who fostered or regulated per share redistribution through domain ordinances. Assuring that all tenants had the maximum opportunity to succeed, to pay rents in full rather than fail, was in the self-interest of landlords. Assuring that tenants did not simply cultivate all of the poorest land in a community, assuring that they had access to good and even excellent land limited the chance of tenant failure.

By the nineteenth century, landlord commitment to *warichi* may have weakened. As discussed for Iwade village, there is evidence that some landlords of the Meiji era and later felt that their own revenues suffered under these arrangements, and they were unhappy about the status quo. They felt that redistribution made it difficult to control tenants and to encourage them to farm assiduously. One early-twentieth-century landlord wrote in frustration: "Since tenants are allocated shares *(kenmae)* from among those of the entire village, it is difficult to record clearly in accont books which tenant is cultivating what land. From the landlord's perspective, that in itself makes managing tenants unprofitable. On the other hand, the weakening of the landlord's management to such an extent is profitable for the tenant who rents the share."[25]

The presumption underlying this landlord's dissatisfaction is that if one could identify inefficient tenants, they could be removed and arrangements made for more assiduous or astute tenants to take their place. The corollary to this presupposition is that either rents took the form of sharecropping, in which the landlord would share in increases in yields from better cultivation, or landlords could raise rents to accord with increased yields. While such presuppositions might arguably be valid much later in the twentieth century, Tokugawa circumstances were considerably different (see below).

Even where there was no tension between landlords and tenants over use of *warichi*, the existence of the system presented several challenges to landlord managers. Since lands were not rented out plot by plot, but share by share, setting rents based on the quality of land would have been a complex challenge. Records would have had to be kept of the rents negotiated between the landlord and each tenant for each type of land in the landlord's portfolio. Because rent would have had to be negotiated for each tenant across the full range of land types, costs in time for contracting would be high for landlords. For tenants, the same kind of complexity and costs would have been introduced—the need to negotiate over each section of land in their portfolio, potentially with more than one landlord.

In fact, Echigo village *warichi* practice eschewed any process of plot-by-plot landlord-tenant rent negotiation for each type of land; instead, rents were set for the entire portfolio as a unit, and, furthermore, rents were not negotiated between each landlord and each tenant but were set for each share of land on a village-by-village basis. There was one rate for renting shares and partial shares in each village. This simplified the process of negotiating rents. It represented a form of collective bargaining between two groups that were analytically distinct— tenants, on the one hand; landlords, on the other—but in fact often overlapped because of the presence of many owner-tenant farmsteads.

One might imagine that the Meiji legal structure ended village-wide rent structures, but, given the landlord's complaint quoted above, clearly this was not the case. Evidence indicates that even after the Meiji Restoration and well into the twentieth century, redistribution of rented lands continued. The old structure of villagewide rent determination kept rents uniform throughout the village and, therefore, quite directly restricted landlords' rent setting.

Even in areas where land rotation was completely abandoned, villagewide determination of rents tended to be the rule.[26] Evidence for this practice can be found at least as late as 1929.[27] Rent calculation took into consideration the national land taxes due, prefectural taxes, and labor and taxes due to the village, as well as rents due the landlord.

Protection from arbitrary rents as well as the benefit of simplified contract negotiation underlay pressures by tenants to determine uniform rates per share of land cultivated instead of individually negotiated, open market rates for each plot based on the productive capacity of the land.

A variety of considerations argue that renters did indeed benefit from villagewide rent determinations. Part of this evidence is found in rental data. Other evidence comes from the pattern of large-scale landlord development in Niigata Prefecture and the comments of early-twentieth-century observers.

First, data on rents presented by a prefectural office report on redistribution, *Niigata-ken ni okeru warichi seido* (The *warichi* system in Niigata), indicates that maximum rents, those charged when there was no crop shortfall, remained stable for long periods of time. The authors present dozens of examples of rents charged in hamlets. These show very few rent changes during the late nineteenth and early twentieth centuries. While the authors indicate that some rents were lowered when there was a crop shortfall, they note very few upward modifications of standard rents. They stress that it was difficult to change the levels of rents charged; for the most part, rents tended to remain constant for decades.[28] When changes are evident, there was usually only one change per hamlet over the years covered by the survey. Since rents were often stated as a fixed volume of grain due the landlord and not as a share of the crop, rents represented a stable, if not absolutely fixed, cost for tenants. Tenants could readily profit from increases in per hectare yields that accompanied the improved farming techniques of the late nineteenth and early twentieth centuries.

Second, deals that landlords made with tenants to eliminate traditional practices suggest that tenants felt protected by villagewide rent determination. When landlords succeeded in convincing tenants to give up redistribution, they literally had to buy tenants off. In one 1919 contract, the landlord not only paid his tenant ¥225 as compensation for the loss of some of his traditional rights, he also specified that rents would be determined by the village as a whole.[29] In this contract, the landlord bought greater flexibility in his choice of tenants: the document specifically stressed the landlord's right to rent the land to another. He also gained some benefit in setting rents. The rental arrangement abolished by this document had given the tenant particularly strong customary rights: rent was relatively low but, unlike in other areas, was not reduced except in the case of very severe crop shortfalls.[30] By selling these rights, the tenant risked some increase in rents and loss of some protection in case of severe crop shortfall, yet he did not thereby give himself over to arbitrary rent assessments. They were to be determined by the village, according to standards with which the tenant was probably already familiar, and they would be lowered in poor harvests.

A third indication of the favorable impact of redistribution for tenants is evident in the contrasts between areas where Niigata's exceptionally large modern landlords developed and in the pattern of landlordism's growth in the *warichi*-dominated areas. Almost all the extremely large late-nineteenth- and twentieth-century landlords were concentrated in Kita Kanbara County, one of the two northernmost counties (*gun*) of the province and one where redistributive practices appear to have been poorly developed if they were present at all.[31] The areas of frequent *warichi* sprouted some large-scale landlords, but their holdings did not compare in scale with these One Thousand Hectare Landlords (*sen chōbu jinushi*).

Indeed, the pattern of development of large-scale landlords in *warichi* areas is distinct from their northern neighbors. In the north, large-scale landlords acquired much of their land after the Meiji Restoration. In the southern areas of the prefecture, the bulk of land acquisition by very large landholders took place before the Restoration, under conditions that were partly dependent on practices associated with local office holding.

The Imai family of central Echigo, for example, built their substantial landholdings in part through the acquisition of special perquisites granted to Tokugawa era village officials. Village headmen in their district received tax exemptions for two shares of land as office emoluments. These lands were exempted from all taxes, not just land taxes. They also received land as compensation for their offices. Both provided opportunities for them to retain wealth from their farm enterprise and reinvest that wealth acquiring more land, lending to others, and engaging in similar business activities. By the time of the Meiji Restoration, the Imai had accumulated such perquisites in thirty-five villages, and these lands constituted perhaps a quarter of their approximately 500 *chō* of landholdings at this time.[32] Their acquisitions peaked in the 1830s, at which time they held land in more than sixty villages in at least four different baronial domains! Thereafter the pace of Imai acquisitions declined rapidly, and, in the years after the Restoration, their total holdings declined by about 20 percent.[33]

Pre-Meiji landlords like the Imai generally had a difficult transition into the modern era throughout Japan, in Echigo/Niigata, the loss of tax exemptions associated with their duties as local administrators was a particularly significant factor in their being superseded by the One Thousand Hectare landlords.[34] Their disappearance was one more challenge to landlords associated with the new economic order based on a rapidly growing cash economy and more fully capitalistic economic activities.

While the Meiji transition cost large landlords office perks that fostered land aquisition, for tenants and small holders, joint ownership's protective mechanisms remained: villages and/or tenants continued to practice both allocation of cultivation rights based on shares incorporating the full range of village lands, and villagewide determination of rents. For smaller holders in the early Meiji era, the continuation of *warichi* and its rent-determining practices placed some limits on extensive large-scale landholding and limited tenancy.

That opportunities for small holders to survive were better in *warichi* areas than non-*warichi* areas is suggested by evidence on tenancy rates. While accumulation of land rights was possible, *warichi* limited the extent of land acquisition by protecting small holder access to good quality lands as part of a broadly risk/opportunity–diversified portfolio and stable rents. In *warichi* areas tenancy rates were lower than for the non-*warichi* areas in northern Niigata. In the early Meiji era, rates in Kita Kanbara were 68 percent. In contrast, they were only 40 percent in Nishi Kanbara, where redistribution was practiced along with villagewide determination of rents.[35]

Fourth, early-twentieth-century observers tell of ardent tenant attachment to redistributional traditions. In addition to the comments of the landlord quoted above and the practice of buying out traditional tenant rights under *warichi*, the authors of *Yamamoto son shi* (A history of Yamamoto village), writing sometime before 1929, concluded their discussion of tenancy with the following remark: "There are arguments that [the traditional system] should be abolished, but they have encountered the unbending opposition of tenants."[36] Prefectural officials who wrote a detailed description of redistribution practices published in 1929 also noted that tenant opposition was one reason for long-standing continuation of redistribution.[37] Finally, the fact that in many villages traditional practices continued only on rented land suggests their strong appeal to tenants.[38] In light of this evidence, the cooperation of landlords in creating a joint ownership regime might not have been absolutely essential. In the late nineteenth and early twentieth centuries, despite the desires of landlords, tenants conducted redistributions among themselves.

It is tempting to ask if there was a relationship between beneficial tenant involvement in redistribution and early-twentieth-century tenant unionization in Niigata. By tradition, tenants were engaged in negotiations to set rents at each redistribution. They also worked together along with landowners in determining the allocation of various tax and labor obligations. Such experiences gave tenants a sense of common interest with each other and a glimpse of what common action might accomplish. Could these experiences have made it easier to form tenant unions in Niigata?

While investigation of this possibility is outside the scope of this study, there is direct evidence that redistribution played a role in the ultimate act of successful tenant union efforts to increase their own landholdings. In 1922, twelve tenants in Yorobigata organized a union in order to improve their economic condition. By 1937, they had bought out the last three remaining large landlords in the village. As the final act in converting tenanted land to private holdings, they conducted one last redistribution to divide their new acquisitions among themselves.[39] Local traditions provided an equitable division of the fruits of their common labor and thus facilitated cooperation among the union members.

## Conclusion

This chapter has focused heavily on early-nineteenth-century and post-Restoration evidence to discuss the position of landlords and tenants as well as the malleability and durability of redistribution practices long past the Meiji Land Tax Reforms that theoretically eliminated joint landholding practices. Per share proportional *warichi* practices had considerable appeal beyond the full shareholders who dominated village society. Tenants, too, saw benefit in the practice; they continued to use it among themselves even when hamlets formally abandoned the system, and it provided a foundation for joint negotiations over rents, both before and after the Restoration. Further, the system was adapted in some cases specifically to provide cultivating opportunities for the landless and small shareholders. In addition to adapting to microclimatic change and cash crops, as discussed in Chapter 7, some villages completely renegotiated the degree to which they applied redistribution practices to their main arable lands, discontinuing them for newly reclaimed land—an effort to maintain incentives to keep as much land under cultivation as possible and to adapt to changing circumstances. As in the Iwade village evidence, the tension between private interests and broader community interests, and between self-interest and altruism, created continuous reevaluation and adjustment of joint ownership practice before and after the Restoration.

Perhaps most impressive, however, is the degree to which shareholders and tenants purposefully ignored both the letter and the spirit of Meiji law. They created polite fictions of compliance and continued to use those customs that they felt were most beneficial to their livelihoods. Although affecting fewer and fewer fields over the course of the twentieth century, this mentality continued to be reflected in some cases down to the 1970s. By that time modern techniques for limiting natural hazards and compensating for natural differences in productivity of the land, the diminished role of agriculture in the villager livelihoods, and the state's exercise of eminent domain combined to render redistribution practices untenable. Though abandoning joint ownership in the long run, shareholders and tenants did not respond to Meiji ordinances robotically. Rather, they considered their own values and customs, often thoughtfully adapting landholding practices to the new regime, the increasing commercialization of agriculture, and the rise of new economic opportunities.

## 9

# Final Reflections

The preceding analyses raise a number of interesting perspectives about Japanese history, the functioning of joint ownership in early modern and even twentieth-century Japan, and the implications of that history for understanding economic activity and motivations in human society. My analysis has tested critically postulates drawn from the scholarly literature on joint ownership in Japan, developed a classification of major variants, and revealed a number of unusual wrinkles in the actual practice of the most common form, per share redistribution. The implications of these analyses call into question common ways of thinking about the relationship between ownership rights and economic incentives. In this final chapter I pull together findings and implications drawn from the preceding analyses. Before a fuller elaboration, the key points are as follows:

1. A significant implication of divided political authority in early modern Japan was that many policies affecting villagers developed first at the village level and secondarily at the level of daimyo administration, resulting in a diverse array of practices, as the case of land rights illustrates.
2. The presence of joint ownership regimes did not substantially deprive cultivators, villagers, or domains of the incentives to invest in either structural improvements (e.g., expansion of arable or irrigation infrastructure) to or maintenance (e.g., paddy

189

pans, application of fertilizer) of farmlands; joint ownership successfully adapted to diverse, changing economic and political circumstances.

3. Joint ownership did not result in a "tragedy of the commons" and lead to resource degradation.

4. Consequently, we may ask if the contrasting economic outcomes of joint ownership and private ownership have been overdrawn.

5. Further, joint ownership patterns frequently persisted and villagers adapted them to new geographic and social conditions because they served purposes valued by cultivators.

6. Joint ownership regimes were not just the outcome of high natural hazard risk; some types had completely unrelated functions, and while others had links to natural hazard risk, that alone was not a sufficient condition to implement joint landownership.

7. In addition, variation in joint ownership practice and its presence or absence in different villages facing similar natural hazard risk must be explained as the result of available local leadership/managerial skills, economic altruism, and a history of successful cooperation.

8. Through joint ownership systems, cultivators expressed a sense of parity, equity, and/or fair shares as important community values toward land.

9. While valuing land as an economic resource, early modern landownership practice often did not treat it as simply a commodity but as one element embedded in a variety of economic, political, and social networks.

## Shared Power, Diverse Practice

The preceding analyses demonstrate that one cannot get a sense of transformations in landownership rights across Japan without conducting regional and local studies and incorporating them into an understanding that accounts for minority as well as majority practices. The scale of minority practice detailed here reveals that emphasis on just the largest trends leaves out too much. At minimum, a substantial minority of Japan, up to a third of Honshu, Shikoku, and Kyushu practiced some form of joint village control of arable land.[1] Even within joint landownership, three patterns existed—per capita, per family, and per share.

In the context of Japanese history, awareness of multiple land-ownership structures directs attention away from the political center to lower levels of administrative action. Even if we assume that land surveys were routinely implemented according to the models historians frequently ascribe to Hideyoshi and his Tokugawa successors, they did not permanently establish either a uniform system of land tenures or a universal pattern of nearly modern private ownership. At absolute minimum, the rise of village-based *warichi* immediately after land surveys indicates that real control of land tenures lay at much lower levels than that of the emerging hegemons. Other evidence also indicates that control of land tenures lay in the hands of daimyo and/or villagers. Above the village level, some domains (like Satsuma) successfully controlled land tenures, but even they could have trouble, as the case of Tōdō domain indicates. Further, even within a domain, the structure of land-ownership rights could vary. Tōdō domain represents one example, but within lands directly controlled by the shogun (for example, as noted in *Jikata hanrei roku*), ownership rights were not uniform; some villages practiced joint ownership of land, others did not. Indeed, in the later nineteenth century, even the Meiji government had difficulties implementing private landownership. I have noted that many Echigo villages found their own devices for circumventing the principles that underlay the distribution of Meiji landownership certificates. Even when officials knew these practices continued, they simply acquiesced to local custom in many instances.

While *warichi* may not represent a majority of tenurial regimes in early modern Japan, the breadth of its presence speaks persuasively about the limitations of political power at both the national and even the domain level. If Hideyoshi and his followers sought to impose a nationally uniform system of land tenures, it clearly failed. Even at the domain *(han)* level, village initiative tended to dominate. This was clear in the development of Kaga domain's *warichi* practices. When domain authorities sought to encourage the practice, they left most decisions in the hands of villagers, attempting only to license those who made measurements (to assure reasonable techniques were followed) and to set a maximum interval between redistributions. The case of Tōdō domain's failed effort to implement *warichi* drives home the point.

That diverse tenurial regimes survived the major land surveys and coexisted should come as no surprise; ruling classes had no great incentive to invest the man-hours necessary to track ownership meaningfully. Economists have long understood the substantial up-front costs of establishing property rights and institutions, and that for such an investment to be worthwhile, benefits needed to outweigh costs.[2] It would have been hard for any ambitious ruler in the early seventeenth century to make that case. Nationwide, creating and administering a uniform system of tenures would have meant first establishing a true, centralized system of administration, an effort that would have directly placed the Tokugawa—a large, but still minority landholding barony— in direct conflict with the jealously guarded traditional rights of several hundred daimyo. Even Hideyoshi, who had personally conquered large parts of Japan, did not feel strong enough to take such steps. After the Osaka campaigns and the death of Tokugawa Ieyasu, there was substantial uncertainty that the new order would hold together, and the fumbling initial response to the rebellion in Shimabara (1637) bears out the tenuous ability of the shogun to treat daimyo as military, much less bureaucratic subordinates. Within domains, control was no more certain. The middle to late seventeenth centuries were marked by so-called house disturbances (*ōie sōdō*), and these factional tensions often rent domains in bloody conflict.[3] Settling such political-military challenges took precedence over clearly defining the landholdings of individual families, creating a system of private landholding, and developing the survey tools that would make boundary definition workable.

In an age of scarce administrative resources, limited communications technologies, and pressing issues of establishing domestic tranquility and maintaining an emergent, as yet fragile, military-political balance of power, rulers could not afford investments to establish and track private ownership. Since the land tax and other taxes on villages were assessed on and collected from villages as corporate units, not from individuals, the administrative expenses of the daimyo and the shogun (the largest daimyo) were kept in check. Villages created the structures for and bore the costs of collecting taxes from individual households. This arrangement allowed rulers to devote their resources primarily to maintaining public order and other government functions without tracking individual landownership.

Arguments that Hideyoshi's surveys tied landholders securely to specific pieces of land and secured for them the fruits of investments in improvements misrepresent the mechanism by which increased security of tenure was achieved. Village practice was responsible, and, as indicated below, its guarantees were incomplete. Echigo village practices protecting long-term investments in lacquer tree production provides a concrete example of village efforts to structure land rights so as to foster individual investments. However, when land rights were contested, internal disputes were generally handled first and foremost within the village. As a rule, domains adjudicated disputes over land that involved multiple villages, not those in which the matter was strictly internal, as seen in the disputes introduced in Chapter 7.

## Economic Incentives

In the preceding discussion the issue of diversity in landownership rights has theoretical as well as methodological implications. As noted at the beginning of this book, one system of ownership rights often does not simply replace another. Diverse and diffuse control of land tenure rights forces us to recognize the simultaneous presence of multiple land rights structures as we consider broad trends and their economic outcomes. The cases described above have explored nuanced gradations between public and private ownership in both the early modern and modern eras, and demonstrated that some of the joint ownership structures seen as inefficient or outmoded exhibited a sensitivity to investment incentives and surprising persistence, even in socioeconomic settings far different from those in which they originated.

Early modern farmers and domains employed multiple structures of landownership rights in different contexts at the same time, and to some degree that practice continued even in the twentieth century. Something akin to fee simple ownership was a common pattern of arable landownership in Tokugawa villages, but even in these villages it was not considered suitable for managing the resources of the commons. The latter resources were treated as essential for all farmers in the short and intermediate term, and, to protect the interests of all, these lands were subject to joint ownership by shareholders of one or more villages. Joint ownership was also applied to a significant minority of arable lands in specific circumstances. Our typology stressed three key outcomes: maintenance of a labor force, provision of a scarce resource (land for vegetables), and some element of sharing geographic risk and diversity. Often, all these ownership permutations coexisted in a domain or even in a village; they were not mutually exclusive at these administrative levels, let alone nationally.

At least superficially, the presence of joint forms of landholding raises the possibility that the degree of security of tenure that existed in many parts of Japan has been overstated. All three major forms of joint land tenure cut any direct ties between an owner and specific plots of land. All three involved nonmarket rotation of some sort, off one piece of property onto another, and potentially robbed cultivators of the fruits of their investment.

However, evidence presented above suggests that we cannot simply conclude that joint ownership compromised security of tenure, and, further, that the widespread contrast drawn between the effects of joint versus private ownership has been overdrawn. These systems could function or be designed in ways that protected incentives for both villages, acting as a corporate unit, and individuals to invest in the land. It is important to separate village and individual incentives; they do not always bear on the same activities or coincide. At the lowest level, villagers in the most common joint ownership regimes (per share) used their capacity for self-governance to craft institutions that showed considerable sensitivity to maintenance of economic incentives for individuals to invest in both expansion and improvement of land, and to invest in crops with intermediate investment horizons. Individual land reclamation was rewarded with an increase in the shares a villager held and/ or exemption from redistributions shortly after land was brought under the plow. Provision was made to protect intermediate-term investment required to produce lacquer trees for market. In these ways villagers demonstrated considerable flexibility, adjusting to local and changing circumstances, a pattern explored in villages in the modern Tokamachi area. Nothing in the structure of joint ownership dissuaded cultivators from investing in weeding, fertilizing, or making other changes to cultivation that generated immediate improvements in crop yield. By their behavior, villages demonstrated through the ways they structured joint ownership that it need not be economically detrimental, that it was not, ipso facto, a constraint on economic growth and change or antithetical to economic modernization. Villagers implemented policies commensurate with the growth of agriculture.

Similarly, nothing in the structure of joint ownership dissuaded villages as corporate entities from investing in improvements to land. These investments took two forms, expansion of arable through joint reclamation, for which domains offered tax incentives, and improvement and maintenance of irrigation and drainage works. In the latter case, domains typically offered some measure of support, at least for regional cooperative irrigation and flood control works. The interests of domain coffers and the economic well-being of villages overlapped in the realm of investments in land.

The presence of joint ownership does not challenge the story of low-level change in economic behavior that laid the foundation for Japan's modern economic transformation (as T. C. Smith and others have contended); however, its presence does mean that we must envision a more subtle and diverse set of mechanisms through which that story unfolds. Echigo villages exhibited a number of the developments that Smith pointed to as evidence of increased investment in land and market engagement. I have noted in particular the presence of agricultural treatises, crop diversification, flood control, and the like; I also noted changes in family structure that typically accompanied growth of day labor markets in villages (Chapter 2). It would be ideal if there were statistical evidence in addition to the qualitative evidence that agriculture in *warichi* regions improved over time and on a par with other regions. Such evidence does not exist, but changes in land-tax rates from the late Tokugawa to early Meiji provided some indication. These data indicated that Echigo yields as well as area under the plow increased over the period. Cases from Okayama indicated the same claim can be made for per family joint ownership.

Finally, in considering the viability of the dyad that joint ownership equals economic disincentive and private ownership equals economic incentive, it is worth repeating Miller's conclusion that the modern corporation, for all its vaunted efficiency, cannot escape widespread internal free rider problems; there is no simple way that self-interest can be marshaled to eliminate unproductive behaviors just by getting incentives right. In addition, recall Barzel's argument that no form of property captures for an owner all elements of ownership, that legal systems capture only a limited range of rights associated with ownership, and that typically there are multiple owners, even under what is typically subsumed under the classification of fee simple possession.[4] In practice, neither joint ownership nor private ownership lives up to its caricature. Free rider problems exist in both regimes, and, in both, the advantages of cooperation can outweigh their costs.

Villagers also adapted these systems to changing socioeconomic circumstances to a considerable degree. Joint landholding frequently survived even after the Meiji Land Tax Reforms. Evidence for the persistence of *warichi* in Echigo (and other areas) long after the Meiji government's privatization efforts suggests that in a number of instances these tenure systems had very strong internal roots and resilient adaptive capacities. I have focused heavily on per share proportional redistribution, the most widespread form of joint ownership of arable land, particularly in the Echigo/Niigata area. However, other forms of redistribution also survived past the Meiji Land Tax Reforms: per capita allocation in Satsuma (to which we can add Kudakajima, Okinawa) and equal allocation in Seiriki hamlet, Kumayama, Okayama.[5] These adaptations and survivals took place despite efforts of a central, external authority with clearly coercive powers to break them down. They persisted in tandem with the growth of a cash economy, industrial capitalism, and scientific/mechanical advances in agriculture and gradual diminution of the role of agriculture in generating family income. The strength of joint ownership practices in the face of an increasingly effective and often repressive Meiji and prewar state is itself remarkable, but, equally notable, these customs also survived the land reforms of the American Occupation and continued into the 1970s—developments that bespeak their centrality to village agriculture in some locations. While Americans may feel that only practices sanctioned by law are secure, it is clear that joint land tenure practices were respected by local residents and their governments to the last. The decision by cultivators to abandon joint ownership in the postwar era was made as the economic foundations of Japanese life changed during the era of rapid economic growth or, as was true in Kumayama and other areas, as jointly owned lands were lost to state exercise of eminent domain.[6]

Village-based examples were especially self-sustaining, supported by internal dynamics. The patterns of joint control and the many cases of joint village management of land both argue that compulsion, externally imposed (Hardin's argument), was not necessary to design, maintain, and adapt these tenurial systems. Internal structuring of incentives provided the carrot, supplemented by group pressures if sticks were needed. In combination, they limited potential free riding so that, as in the modern corporation that Miller describes, costs were outweighed by the gains that these arrangements brought participants. Early Meiji land-tax and qualitative data support this contention.

## Joint Ownership and the "Tragedy of the Commons"

What circumstances operated to prevent a "failure of the commons," a severe degradation of the arable land resource that thinkers like Hardin predicted?

Two clusters of circumstances mitigated against a tragedy of the commons. One set, especially operative in the Tokugawa era, left cultivators little option but to consistently maintain, renew, and extend the land they farmed. A second set, operative in the post-Restoration years as well as during the early modern era, limited pressures to overexploit the land.

Given the core crops that were typically planted, the failure to invest in maintaining the land would have amounted to shooting oneself in the foot. A poorly maintained paddy bed reduced water retention and immediately reduced per hectare yields. Failure to fertilize seedbeds and crops, particularly some vegetables, also quickly reduced per hectare yields. The fact that some nineteenth- and twentieth-century landlords raised complaints that individuals failed to maintain their fields well in the single year before a regularly scheduled redistribution reinforces this impression. One could not shirk for any extended period. Landlord complaints must be taken in a context in which farmers tended to be sensitive to quite small differences in the efforts of their tenants and neighbors, but in addition, their impact was readily reversible, and declines typically were not long-term or significant enough to suspend redistributions. Temporary declines in per hectare yields did not threaten the long-term viability of arable land (in contrast to soil depletion in nineteenth-century American tobacco- or cotton-growing regions). Any latitude for free riding occurred in a very narrow band of opportunity, and soils recovered quickly. Furthermore, it is important to recall that many villages implemented redistributions only for cause, a significant flood or landslide, and not periodically. In such villages, redistribution could not be predicted. No one could safely and with assurance plan to reduce investments in the land. This remained true even as population increased and the economy diversified during the early modern and modern eras. However, participants were surely aware that, even with periodic redistribution using a random draw, one might receive the same fields they had farmed heretofore, adding further risk to any attempt at free riding.

The nature of early modern agriculture limited the potential sources of degradation in most *warichi* areas, and farmers did not plant crop varieties that placed a heavy drain on soil nutrients. In rice agriculture, the major nutrients are delivered to plants via irrigation water rather than from the soil directly. Thus rice plants themselves neither depend on the soil for most nutrients nor do they aggressively leach nutrients from the soil. This meant that a key issue in rice agriculture is adequate supply of irrigation water and its regular, predictable distribution. The most commonly grown dry-field crops, either as a second crop on paddy or on fields reserved for dry crops, also required little fertilizer: wheat, barley, millet, barnyard millet, soybeans, giant white radish (daikon), red beans (azuki), mulberry, arrowroot, lacquer trees, and the like. (Other nutrient-intensive crops such as cotton and tobacco grew in importance during the early modern years but still represented a small fraction of acreage.) In the case of some vegetables, fertilizer could certainly boost yields, but failure to provide it led quickly to harvests of lower value and volume—not much chance to free ride. Even when they had commercial value, such crops also tended to be supplemental and produced on limited acreage.

The key problem in rice paddies lay in maintenance of good water access and distribution. Even with a well-controlled supply of water into a paddy, the irrigation gates, paddy ridges, and beds had to be in good repair to assure that one's allotted supply of water would not be wastefully drained and that appropriate amounts would reach all the plants in a field. The essential investment in this endeavor was labor to minimize unwanted drainage and to create an even paddy bed that permitted all plants to get the necessary water. Failure to do either resulted in immediately detectable declines in yield.

The presence of short intervals between redistributions evident in periodic Echigo practice reinforces the preceding explanation. Three-, four-, and five-year intervals were common. It is hard to imagine such frequent periodic redistribution if free riding actually caused significant long-term loss of soil fertility, damage to paddy pans, or facilities for regulating the flow of water into fields.

Pressures to overexploit the land were limited. Although a number of crops found their way into the expanding regional and national trade networks of the seventeenth to nineteenth centuries, external demands for increased production came gradually, were relatively modest, and give no evidence of encouraging the degradation of fields under *warichi*. There were instances in which the effort to extend arable land into upstream fields or to introduce irrigation to convert dry fields to paddy resulted in disputes over water for older downstream fields. Both villagers and officials often resolved such disputes in such a way as to protect older fields. However, extension of cultivation into common lands and their privatization could and did lead to very different outcomes for nonfarm land, in which woods, watershed, and erosion control suffered.[7]

Thus, rural market activity grew significantly and with important consequences over the Tokugawa period, but generally within the confines of a self-sufficient base for most of the rural population. Farmers transported vegetables to towns and cities for sale, they processed soybeans into tofu and other products, the silk and paper trades expanded, and rice not paid as taxes was converted to rice wine *(sake)*, all in increased volume over time. Yet given the need for households to supply all of their own basic food and to pay the majority of land taxes in rice until the 1870s, farmers had limited flexibility to abandon production of these basics in favor of full commercial farming. The necessary food and tax rice simply could not be purchased on the market in sufficient volumes.

At no point did the market pressures from outside a village encourage overexploitation of arable land itself on a wide scale. There was no substantial incentive to overuse farmland. Common pre-Meiji disputes over land use involved agriculturally marginal lands—common lands, woodlands, and the like—should they be maintained as is, or were they better devoted to extending arable land to increase foodstuffs or a growing share of commercial crops? In a number of cases, pressures on these lands resulted in a loss of commons.[8] There was pressure to convert these to arable. In the seventeenth century, the tendency had been to use such newly cultivated lands for foodstuffs; by the late eighteenth century, the trend in some regions of Japan, especially those surrounding major metropolitan centers, was to plant more commercially oriented crops. However, population and commercial pressures did not have a similar, extensive impact on exhaustion of arable land.

For much of the early modern era, limitations on population movement made it difficult to pursue a free-riding strategy. Losses from overexploitation might not be of substantial consequence in contexts where "moving on" or locating alternative careers was an option (e.g., the American South or West in the eighteenth and nineteenth centuries), but an early modern Japanese farmer could not readily pull up stakes and move. In the early seventeenth century, such mobility had been possible. The century of civil war had created such disorder and destruction that a large amount of land was removed from cultivation. During the late sixteenth and early seventeenth centuries, there were substantial efforts to bring these lands under cultivation. Such efforts provided opportunities for a number of people to move to new farmsteads and expand their farm operations, and in some cases provided opportunities for urban and other poor to become full-fledged farmers. But by the late seventeenth century, such opportunities were rapidly declining. Even if one could move to another region, establishing oneself as a farmer in a new community was difficult: credit depended on having collateral in land; newcomers had no land. During much of the early modern era, agricultural day labor markets were expanding but not yet very fluid. Finally, no significant frontier areas existed. With little chance to move one's farmstead, disincentives to over-exploit farmlands were reinforced, effectively forcing those contemplating a free rider strategy to face the ecological or social consequences of their actions.

This particular set of incentives and disincentives changed dramatically in the mid-nineteenth century. The Meiji government removed old constraints on domestic trade, introduced much improved transportation, and placed land-tax obligations directly on the backs of individuals, each of whom was given a formal certificate of landownership. The new land-tax system mandated cash tax payments and made no provision for lowering taxes in years of poor harvests.

This new system permitted substitution of cash crops for subsistence crops, but the shift to full market orientation took time. The basic range of crops, farming techniques, and patterns remained largely unchanged for several decades. This new environment certainly contributed to the abandonment of *warichi*, but change was often halting and some regions continued to practice joint landholding. Even in the post-Meiji and modern eras, I have yet to find either documentary evidence or discussion in oral interviews that joint ownership actually created significant free rider problems.

## Explaining the Persistence of Joint Ownership

Even if we assume that lands under joint ownership maintained yields and that there was no significant ecological damage from redistribution, the question remains: why did the practice continue for so long? Why would a system of joint landholding and periodic reallocation of access to fields work for centuries in such diverse socioeconomic and political environments as seventeenth-century agrarian and twentieth-century industrial Japan? Why would farmers, engaged in a risky business to begin with, apparently compound their uncertainties by allowing a lottery or other nonmarket device to determine which fields they tilled? Although joint ownership declined in some areas, both during the Tokugawa era and particularly after the Meiji Restoration, villagers resisted its abandonment in others; its persistence speaks of a strong attachment to it, one strong enough to repeatedly invest the kind of careful attention and work exhibited by Iwade village shareholders.

Based on the Echigo practice of *warichi*, the following observations on its persistence seem pertinent. First, mechanisms existed through which redistribution outcomes could be attenuated. If parcels allocated for individuals were unsuitable for some reason, they could be exchanged privately, as in Iwade. Second, the system made provision to protect investments that had long-term return horizons. Constraints on innovative agriculture—especially in the early modern era—that were present came from other sources such as irrigation needs in areas where water supplies were limited and had to be rationed, as in the Yoshikawa area case noted in Chapter 7. Third, while it was seldom a major driving force in shaping *warichi*, supravillage society's inertia or even active support reinforced the legitimacy of local practice and the influence of village elites who had a preponderant influence on how the system operated.

Fourth, the system provided tangible benefits for participants. Insurance, welfare, or other benefits were significant enough to offset the need occasionally to swallow one's pride and accept compromises that were not always to one's liking. The continued use of *warichi* by tenants in the post-Restoration decades suggests that cultivators—from those who exclusively or largely cultivated rented lands to those who held full land management rights—benefited from redistribution practices. It did not just serve the interests of one class of cultivator. Tenants benefited as well as principal holders of cultivation rights: on the whole, they saw *warichi* as protecting their farming enterprises.

The system was adaptable to new economic circumstances, too. The adaptability and the malleability of functions evidenced by *warichi* over time reinforces the idea that it provided real value, from insurance to welfare and other functions. Especially after the end of the shogunate, when the system had to be maintained in the face of inimical land-tax and property regimes, its adaptability was impressive.[9] The most outstanding example examined here transformed one Tokamachi area *warichi* practice from a device to recruit labor for one man's agricultural investment into a form of "social security."

Respondents in interviews, when asked why the tradition continued so long in their villages, replied that people in their area generally got along better than folks in other villages. While this is something of a stock reply in "in-group"-conscious Japan, I think that there may be an important element of truth here. That is not to deny the possible role of "effective persuasion," including threats, but ultimately, even when villagers could marshal outside support for individuals to quit, people acceded to the old ways.[10] There are other indications of ten-

sion within villages practicing *warichi,* but if we conceive of harmony
as agreement on the rules by which disputes are resolved, Georg Simmel's perspective on social conflict, then these villages created ways
to resolve conflicts as part of the ownership system while maintaining
it intact.[11] Within this context, villagers managed to adapt their joint
ownership practices to a variety of changing circumstances—conversion of dry land to paddy, the implementation of new national tax and
land laws and changes in the surrounding social and economic context.

Some might argue that a final issue, quite different in nature, also
helps explain the longevity of joint tenures and their ecological stability
by limiting pressure on the land: the absence of chronic warfare such
as that which confronted Western Europe during the early modern period. Charles Tilly, for example, argues that warfare provided the stimulus for increased centralization and state intrusion into local affairs.[12]
Yet in early modern Japan between 1590, when Hideyoshi subdued the
last of his domestic enemies, and the 1860s, when the antishogunal
forces took up arms, there were only four major domestic battles and
two foreign conflicts. All of these ended by 1637.[13]

That the absence of war reduced the sense of urgency felt by domain and shogunal administrators cannot be denied entirely, but that
did not mean an absence of red ink in administrative budgets and at-
tendant pressure to increase revenues. Japan's ruling classes were not
sufficiently creative to prevent peace from interfering with profligate
spending. On the one hand, costs spiraled as upper-level samurai pur-
sued increasingly luxurious consumption patterns. Close to the top of
any list explaining this pattern of expanding consumption sits the mandatory system of alternate attendance *(sankin kōtai)* that forced daimyo,
their families, and many of their retainers to spend alternate years in
the shogun's castle town of Edo, maintaining two separate homes in
Edo and in their domain. Proximity of daimyo and their retainers in
Edo sparked competition in consumption.[14] A similar pattern surely
operated in domain castle towns where samurai were concentrated. On
the other hand, real purchasing power of government and samurai in-
comes declined from the late seventeenth to early nineteenth centuries
as economic growth and diversification resulted in a secular decline in
the value of the rice and soybeans—the principal forms of tax payment
and the currency in which samurai salaries were paid.

The ways in which rulers dealt with these pressures tell us something about where they found expedient responses to financial exigen-
cies.[15] Facing red ink, domain and shogunal administrators attempted

to raise land taxes, exhorted villagers to be more frugal, and accused them of cheating on taxes. However, they were constrained by problems of maintaining administrator knowledge of agriculture and technological limitations on transportation and land measurement.[16]

In the end, the administrative efforts needed to enforce a more efficient system of land taxation were more problematic than fiscal pressure per se. Shogun and daimyo acted as though taxation was not worth the increased administrative costs or those of overcoming endemic village intractability. By the mid-eighteenth century, when all such measures and more failed, they turned to debt cancellation and reducing samurai retainer salaries. The latter tactics had the virtues of being expedient and administratively economical. For farmers, these policies meant that the key long-term pressures they placed on the land were those of (1) paying a relatively stable rate of taxation and (2) providing for their families.

## Redistribution Systems and the Natural Environment

While we cannot completely rule out hazard risk as a stimulus to creation of redistribution systems, it constituted only one consideration. The malleability of *warichi* by itself suggests that any conception of a tight link between joint forms of landholding and amelioration of natural hazard risk needs to be loosened; other evidence supports this argument. From the start, joint forms of landholding had diverse outcomes, encompassing factors other than natural conditions and serving completely different functions. For example, in the *ōaza* of Seiriki (Kumayama), *warichi* was employed during the nineteenth and twentieth centuries to provide only dry field to assure that all had vegetables. In Kudakajima, Okinawa—to briefly introduce one additional, striking example—access to land (all dry field) was provided by a form of *warichi* to provide basic sustenance for women and children while the island's men sailed off to engage in trade, fishing, and maybe piracy. In this latter case in particular, there was no association whatsoever between frequent flooding and the presence of *warichi*. Satsuma's *kadowari* system maximized retention of labor for the domain's purposes. Some villages employed *warichi* initially as a means to attract sufficient labor to bring new land under the plow. Such regimes and approaches were either completely unrelated to amelioration of natural hazard risk or (as in Satsuma) were secondary to other functions.

Even within per share proportional redistribution systems—those that seem to have the closest tie to natural hazard risk—the intervals between redistributions in similar geographic circumstances varied widely and suggest a considerable degree of variation in villagers' collective judgment about the trade-offs between the effort required to administer joint ownership and the ameliorative benefits of doing so. Contrary to expectations that villages in close proximity, sharing similar geographic, economic, and political environments, would have similar periodic redistribution intervals, examination of 110 Echigo villages showed significant variation. This evidence indicates that such divergences expressed different degrees of tolerance for natural hazard risk. Examination of the circumstances of Shindōri and Kamegai villages in particular suggests this, but we also conversely find numerous examples in our maps of widely varied geographic conditions (even in close proximity) that shared common intervals, something we would not expect if there were clear links between natural conditions and redistribution intervals. Elasticity in risk tolerance may be one reason why joint village ownership appeared only in some parts of Japan subject to considerable flood or landslide hazard risk, or great variation in soil or microclimatic differences.

One must be careful not to fall into thinking that all redistribution in joint ownership was periodic. Recall that in many instances (early Kaga, for example), internal measurement, regrading, and reallocation of land took place only after a flood or landslide of sufficient magnitude for villagers to reach a consensus that they needed to redistribute lands. Here, villagers directly linked redistribution to ameliorating the costs of a natural calamity. Nonetheless, the responsiveness of these mechanisms to natural hazard risk bears further investigation.

Further, it is worth stressing that even in per share redistribution systems the function of the system changed over time. Just because *warichi* was tied at its origin to spreading natural hazard risk among all shareholders and/or tenants did not mean that the community continued to use it for that function. While some shifts, for example, spreading the benefits of land reclamation or recultivation and sharing potential loss from floods, were related to the natural geographic circumstances of the land, the shift to something of a social welfare function in the Tokamachi area was not.

## Leadership, Economic Altruism, and the Role of Experience

All of the preceding evidence indicates an attenuated link of joint ownership practice to natural hazard risk, one mediated by very human considerations. Variable risk tolerance, both individual and collective, was surely an important factor, but others can be adduced. Among them, leadership skills must have been prominent, whether they belonged to village officials or village residents. Leadership resources available from community to community varied considerably. Leadership—the ability to provide direction, coordination, and to resolve dissent—is an essential component in deciding whether and how to respond to natural conditions under any circumstance, but creation and maintenance of joint ownership created special demands for such skills.

To generate agreement to initiate joint ownership required significant leadership and organizational skills, especially in village-initiated practices. Someone had to conceptualize a system of joint ownership and the mechanisms for adjusting and reallocating cultivation rights, and then to convince others to go along. A variety of leadership techniques, from logical persuasion to applying social or economic pressures, were required to implement joint ownership. Leadership might reside in an economically or politically dominant figure in the community or a lower-status but verbally persuasive or respected individual capable of engendering cooperation. Given the possibility that a dominant figure in a village could corner the market on the safe lands in a village and thus minimize his own personal exposure to natural hazard risk, willful top-down pressure seems unlikely as a factor in creating joint ownership. The practice might have originated as a way to allocate access to a landlord's tenants, but that would presume that the practice began in villages where land was totally controlled by a small number of families. Even that function would not explain why shares typically came to be traded on the market rather than just rented out. If economic/political dominance were a key operating factor, it would still take considerable leadership to ensure cooperation of all participants. Such talents would not be present in equal measure in throughout Japan's early modern villages, and that variation surely shaped patterns of *warichi* development.[17]

While it is not possible to document the role of leadership in the formation of joint ownership systems, other evidence presented at both the domain and village levels is consistent with such a perception. In Chapters 5, 7, and 8, I detailed the myriad points of negotiation over classification of village land, *hikikuji*, and other elements of the redistribution process. I described changes and adaptations that took place as a result of these negotiations and the effort that villagers took to get incentives to invest in land right. In the case of Satsuma, domain leadership worked well. Whether by force or arms or persuasion, the *kado-wari* system was established early in the domain's history and appears to have been effectively implemented. In stark contrast, the authorities in Tōdō domain failed miserably, despite possessing military force potentially capable of compelling adoption of *warichi* domainwide.

Two pieces of indirect evidence support the role of leadership at the village level, both involving events at the final stages of joint ownership practice. Two interview informants, one in Nagaoka, the other in Kumayama, both related experiences in which local leadership and persuasion played critical roles in shaping the final dissolution. Both involved confiscation by the state of lands on which *warichi* was practiced through exercise of eminent domain. In the Nagaoka case, my informant bragged about his role in getting reluctant neighbors to cooperate rather than fight the takeover, hinting at his use of subtle threats of less desirable outcomes if cooperation were not forthcoming and better financial deals for the group if they cooperated. In the Kumayama case, my informant took the lead in successfully proposing and adopting the final plan for disposal revenues generated by the exercise of eminent domain.

My Kumayama informant hinted at still a third set of considerations bearing on successful use of joint ownership when he pointed to the fundamental sociability of the Seiriki community. Already noted in a different context above, I raise this consideration here because the experiments of economic psychologists provide evidence that this is an important insight for understanding cooperation generally.

The economic psychology literature on altruism exhibits a consensus on a number of processes that create a culture of altruism. In the context of economics, "altruism" refers to what economists consider "the analytically uncomfortable...fact [that]...a higher degree of cooperation takes place than can be explained as a merely a pragmatic strategy for egoistic man."[18] As construed by economists, this is not a highly idealized, beneficial value; it is a set of residual factors creating cooperation that core economic analysis cannot explain. So while this label might, to Japan specialists, seem to resonate with highly idealized positive evaluations of the village cooperative *(kyōdōtai)*, that view would not be accurate. (As an added measure of social conflict in Echigo villages compared to other regions of Japan, note that James White's study of peasant protests classifies Echigo as medium high in total magnitude of social conflict and medium high in increase in social conflict during the Tokugawa era. His social conflict category includes, but is not limited to, internal village protests.)[19]

To explore this unexpected degree of cooperation, economic psychologists designed small-group games in which participants have the option to behave selfishly and free ride on the others in the group, to behave in a way that is completely unselfish, or to take some option in between. A variety of differently structured experiments have been conducted—single iterations, repeated iterations, experiments with communication among group members and without communication, and so forth—that point out critical elements of small-group behavior. First, in single-iteration experiments as well as the first iterations of repeated-play experimental games, the cooperative, altruistic option is the dominant one among participants; where there is repeated play and communication among players, participants tend to cooperate. Second, cooperation tends to continue until it becomes clear that some participants are taking advantage of the others. Third, cooperation is greater when participants believe their group is benefiting or when "altruistic" behavior is reciprocated. Fourth, if given the option, cooperators will selectively interact with each other. Fifth, a variety of experiments suggest that conceptions of fairness enter into people's economic decisions. Finally, in games with a known end, cooperation fails in the final round.[20] Some of these insights may seem self-evident to readers; many may feel that they are reasonable based on personal experience, but they bear further reflection.

Economic psychologists' experiments suggest that whatever the initial motivation for establishing joint ownership regimes in a village, initial successful experience was critical as a foundation for continued use of this form of landholding. They also suggest that both a sense of community benefit and regular communication—the sensibility suggested by my Kumayama informant—create an ethos that generally supported continuation of joint ownership practice, regardless of whether or not it was by some academic standard an optimal way to maximize each individual's personal return. In thinking about national comparisons between widely separated regions subject to high flood and landslide risk and the presence or absence of joint ownership, it is conceivable that initial experiments with joint ownership were tried, but individual behaviors soon undermined them. Leadership skills and cooperative impulses may have failed to overcome "egoistic behavior."

Once again, conditions of long-term cooperation in maintaining village-based joint ownership do not preclude tensions among the joint owners nor do they preclude a breakdown in the system over those tensions. The discussion above indicates that tensions could be quite common, but in confronting them, the successful joint owners found ways to resolve and accommodate concerns. Such flexibility is hinted at by unexplained elements of *hikikuji* exemptions in the history of *warichi* in Iwade village. However, early and continued successful experience with locally generated (not imposed by a domain) joint ownership must have been an important component in the establishment and maintenance of these landownership regimes. Success in reaching solutions to problems as they arose likely generated confidence in members about their own abilities to handle such situations (history matters, even if we call it path dependence!).

The economic psychology experiments indicating the importance of a group's successful experiences with altruistic behavior and early maintenance of reciprocal altruistic behavior also suggest that, over time, a sense of fairness based on the joint landownership system became ingrained in people's expectations. Successful social arrangements and institutions generate expectations and values that motivate and shape behavior over time. The tenacity with which tenants continued to reallocate cultivation rights in early-twentieth-century Niigata and to fight efforts to end the *warichi* system in Iwade village early and elsewhere suggests just such a development.

It is important to consider "altruism" and the way in which people respond to it because, in the absence of direct evidence from the period when joint ownership regimes originated and lack of direct testimony on how people discussed the details of implementation, scholars have not discussed the social processes by which local institutions were created, and they have implicitly treated the beginnings of joint ownership as a somewhat mechanistic response to natural risks. Instead, I am arguing here that *warichi* practices developed from multiple small-group decision-making processes, processes with dynamics that lead people confronting similar challenges to create different solutions. In the end, explanations scholars have offered previously do not help us understand why some parts of Japan adopted these regimes and others in similar circumstances did not. The experiments of economic psychologists remind us that small-group dynamics are historically significant at the local level, not just in the interactions of those who occupy the national stage. The importance of these interactions remains even when the participants are endowed with different social, economic, and cultural capital. They suggest that my Kumayama informant was actually on to something when he claimed that Seiriki *warichi* worked because of the nature of hamlet society.

## Parity, Equity, Fair Shares, and Ties to the Land

The import of a sense of reciprocity to successful maintenance of joint ownership relates to farmers' sense of fair play. Although I have not treated it as a separate issue, the discussion in this book of different types of joint control and their implementation conveys a sense of what villagers—even tenants—thought of as equitable and appropriate behavior. Joint ownership procedures embody a sense of equity and fairness, especially at the village level, that participants found useful in creating successful experiences with these forms of ownership. We learn something about villagers' central values as we reflect on their behavior managing joint ownership. This is particularly important in light of a rather stereotyped view of how villagers viewed land. The concept of attachment to the land, *aichakushin*, is well known in Japanese anthropological circles, and it is usually thought of as an attachment to the fields one's family has farmed for generations. Nonetheless, in regions where some version of joint land management dominated, such a perspective could not have been a part of villagers' mentality. Attachment to place, to the degree it existed, had to be to a village, not to specific plots of land. In *warichi* regions, villagers expressed a sense of parity or "fair shares" directly in land tenure practices; it is remarkable that they did so especially on paddy land where scholars presume a particularly strong attachment to the land.

With regard to equitable practice of joint ownership, we can identify the following attitudes and values:

(1) Random assignment to fields is a persistent, but not universal, principle of design in allotment systems, a principle that limits the potential for manipulation of the process for personal gain. That one's field allocation was a matter of "fate" rather than human manipulation made outcomes tolerably acceptable as fair. In a very few instances, a fixed cluster of fields was assigned to individual households by lottery rather than assigning each individual field by lottery. While fixed sequential rotation was occasionally employed, the use of field assignment by lot overwhelmingly predominated.

(2) In some programs, villagers divided land areas inequitably for distribution; however, differences in area are considered by modern practitioners with whom I spoke to be misleading grounds on which to base a judgment of inequity. Documentary evidence reinforced this idea. Villagers judged that some land, even within a small and restricted section of a village, produced a higher yield than other land, and the

area of each plot was adjusted to compensate for different soil productivity so that each allotment produced a comparable total yield.

(3) A willingness to tolerate marginal inequities that might appear in any particular distribution correlated closely with an intermediate-term view that each participant had an equal chance to benefit from the same inequity at the next rotation. In the course of a redistribution, land area and productivity were not always carefully measured but estimated by sight. My interviews show this to be particularly true for lands associated with shifting/swidden forms of cultivation, but it has also been found in discussion with people who have cultivated long-standing dry field and paddy and borne out in the coarseness of actual land measurement that appears in Iwade village documents. In these instances, some participants doubted that they achieved full equity, but no one contested the results. Confident that the pattern of field redistribution would be continued over a long time and that all participants, through the lottery, had a chance of receiving those fields that might be considered marginally more advantageous, participants abstained from blocking the outcomes of a redistribution.

(4) Where most village land was involved in redistribution, per share proportionality (the most common type of redistribution) rather than equality in land distribution comprised the fundamental operating principle. This provided the desired communal benefits with minimal sacrifice of opportunity for personal gain.

(5) Per share proportionality led to parity in sharing the natural risks of farming (flooding, poor drainage, landslides, and so forth) and the benefits of joint land reclamation, not provision of basic sustenance for each farm family. In areas employing per share allocation, redistribution also served as a means of allocating the land tax burden among villagers under the early modern system of joint village responsibility for land-tax payment.

(6) In instances of equal per family division examined to date, only a part of the village's land was involved; private landholdings remained undisturbed by reallotment (for example, in Kumayama-machi). The opportunity to expand one's holdings in the *warichi* area was restricted but not the opportunity to improve one's position in non-*warichi* land. This practice also limited the access of outsiders who bought land in the community. New residents were generally deprived of the right to participate until they had lived in the village for a number of years, if they were allowed to participate at all.

(7) Where equal per family allocation was involved, the primary

functions underlying redistribution appear to have been twofold. Equal division provided an incentive for broad participation in farming projects that could not be accomplished by small farm households alone and where monitoring of participant inputs was difficult; such a procedure was sometimes employed in land reclamation.[21] Burning of mountain land for swidden in Okayama is another example. Controlling the burn created a major problem requiring the assistance of a number of people. In all of the instances of which I am aware, local residents who wanted to farm upland areas were recruited to clear brush through burning and then were allocated lands by lot to cultivate. They typically grew red beans, daikon, and other crops that did well in poor soil. They were permitted access to these fields for several years, after which all were planted as timber farms (swidden agriculture also breathed its last in the mountain areas of Okayama and Tottori prefectures in the 1970s). In these circumstances, it appears to have been easier to divide the land equally among participants than to try to monitor their respective labor inputs and reward them proportionally for their efforts.

In other instances, the function of equal per family shares may well have been to provide a minimal primary supply of some basic agricultural good for all those recognized as full members of the village. In these instances, the land provided a limited economic safety net for all farmers in a village or a designated portion of them (for example, in Kumayama-machi, Tōkamachi-shi, and wartime Kōchi).

(8) In a relatively small number of cases, lands were distributed to families on a per capita basis, encompassing each family's size, age, and gender composition. This was the case in the *kadowari* system of Satsuma, in which lands were allocated based on the number of adult males in a family. In the case of Kudakajima, Okinawa, villages allocated lands according to the number of children and women in the family. In these instances, the apparent motivation was providing basic sustenance in relation to the number of family members who worked the land or drew sustenance from it. In the case of Satsuma, it also assured a good supply of labor for other domain and community purposes.[22]

The preceding discussion suggests general tendencies, not absolute conditions. The best cautionary note on this point comes from the discussion in Chapter 7 of *hikikuji* in Iwade village. While explanations were found for a portion of that practice, there were also instances that eluded understanding; these instances suggest special privilege of some sort, privilege that compromised a sense of fairness in some measure. At least some of the disputes discussed in this book also suggest that villagers might express considerable differences of opinion over what constituted fair and appropriate practice despite the tendencies outlined above. Even though the documentation of disputes is slim, it would be a mistake to assume that villagers did not need to resolve differences of opinion or that they did not engage in heated debates during the process. It is hard to imagine the significant restructuring of *wari* that Iwade village saw without considerable, perhaps fervent, villager discussions. Villagers worked at creating a system they could live with and from which they benefited.

## Valuing Land

The presence of joint ownership raises the broader question of how villagers valued land, how it fit into their perspective of the world and their place in it.

In the examination of *warichi* in Iwade village, one practice contradicted expectations regarding the kinds of land most central to villager concerns: the Iwade shareholders' attachment to and care in the redistribution process focused heavily on dry-field sections of the village, not the paddy that observers typically treat as the more valuable class of arable land. Dry-field sections of the village were the most subject to restructuring, and special exemptions were disproportionately exercised on these lands. In the absence of special commercial crops, I suggested that this emphasis came from a concern for basic food supplies (although these, too, might have some commercial value).

More broadly, land was an economic resource, and access to it was critical for the survival of rural families, but to say so is not to say it functioned simply as an exploitable family resource. Especially in the late sixteenth and early seventeenth centuries, one did not buy land in the expectation that it would be "flipped" at a higher price later. Land sales and purchases became increasingly common over the course of the Tokugawa era, but even in the mid-nineteenth century land was not largely viewed as a trading vehicle in itself. Possession of land meant status within a village and maybe even in district or regional society. It meant access to local power in decision making and perhaps even low-level domain councils.[23] For most of the period and in most of rural Japan, land was a point of entry into agricultural livelihoods that inherently involved cooperative networks in planting, harvesting, and irrigation. Recall the Echigo lawsuit in which a nonresident sought to participate in the village system of joint ownership. These various networks provided services not available on the market. One gained access to them through a variety of local personal connections. Over the Tokugawa era, and especially from the late eighteenth and early nineteenth century, markets expanded and increasingly incorporated the demands and preferences of people more and more distant from rural areas, creating pressures to treat land as more of an investment resource on its own, but that change of attitude would not be fully manifested until well into the Meiji era and beyond.

*Warichi* systems represent very directly and unambiguously this sense that land is embedded in social networks and relationships that functioned effectively in small, local markets that were relatively insulated from both the opportunities and the pressures of distant market demand. When they survived in the twentieth century, it was in contexts where market demand did not overwhelm the importance of long-standing local social networks or agricultural use of the land, but complemented them instead. They functioned in an economic environment in which agriculture gradually provided less and less of a share of family income in rural communities, and joint ownership could be made compatible with nonagricultural pursuits. They also functioned in an environment that might change the functions of redistribution, as in Tōkamachi.

Two common practices in land rental and sale arrangements present particularly strong evidence that nonmarket institutional considerations mattered in early modern Japan and that market considerations did not completely dominate people's ties to the land. The first, related to village determination of rents in *warichi* districts, was the practice by which tenants acquired permanent rights to cultivate land, *eikosaku* (it also had other names). This class of tenant, common in the Echigo region but widely found throughout Japan, could not be removed from the land. Such a tenant's rights could in some instances be bought and sold, but if a new landlord purchased rights to the land, the tenant could not be removed from the land. This form of tenure, like *warichi*, survived the Meiji Restoration and continued into the twentieth century.[24]

A second practice, increasingly evident in eighteenth-century Echigo but again present in many other parts of Japan, is the practice of not recognizing sales of rights in land as final. Deeds of sale might conclude with expressions that, because the owner had failed to repay a debt on which land rights were pledged as security, the land was now transferred in perpetuity to the lender; however, in fact, the original owner was still recognized as having a prior claim on the land. Through a practice of *kaimodoshi*, or buying back, the land had to be returned to the original owner, even more than ten years after supposed forfeiture, if the original owner repaid the principal. Both village and domain authorities recognized this right on a widespread basis. Over the eighteenth and early nineteenth centuries, documents conveying land to a new owner bore increasingly detailed statements that the original holder was indeed permanently transferring the land and that in the future neither that owner nor his or her descendents had a claim on the land. All this was to no avail, as *kaimodoshi* continued to operate in practice. Under this regime, ownership was viewed as divided, with the original owner retaining rights to the land, much as a modern economist Yoram Barzel has described for various contemporary forms of ownership.[25]

There were widespread practices associated with rights in land that contest twenty-first-century images of private ownership. These practices are not likely to fit any arrangement imagined by Heilbronner when he defined economic society as one in which consumers act to the greatest extent on economic values. (To be fair, Heilbronner would probably not have imagined "flipping" real estate as a normal part of economic society either.)[26] Nonetheless, given the incentives discussed above, it would be hard to conclude that these ownership structures were necessarily economically irrational.

The discussion above is not meant to imply an idealized harmonious communitarian order (the vaunted *kyōdotai*). I have examined evidence of tensions, evidence that I believe to be simply the tip of the iceberg. As noted in Chapter 2, Echigo villages were far too contentious, even litigious and subject to social protest, to constitute realization of an idyllic ideal. (If one defines community as agreement on the rules of combat rather than as a harmonious social entity, we could still talk of a collective, but such a definition implicitly recognizes the presence of conflict.)[27]

At least early in the Tokugawa era, the embedding of economic relations in social networks and collaborative endeavors reflected the reality of village life in which both the market and the state could not yet provide essential elements (irrigation, labor, and so on) necessary for agricultural production. As the conditions of agriculture and the market changed, local social institutions, joint ownership systems among them, were capable of change and adaptation. The late nineteenth century brought major changes in this regard. The uses of land and the demand for land changed, diversifying its potential uses and leading people increasingly to value it as a commodity subject to expanded market forces. While large landholding in a community may still have conferred a significant sense of prestige and social standing on a local resident, that element of value was lost for outside individuals or business investors, people divorced from the community who increasingly had deeper pockets than local residents. To these market stresses can be added the interests of a modernizing state, which was increasingly willing to exercise eminent domain to create public works or to promote field rationalization, often the coup de grace for joint ownership remnants in the 1970s.

Joint ownership systems expressed a careful balance of private and public good in the context of the particular challenges and opportunities faced by agriculture in different communities over the early modern and even modern eras. The discussion above suggests that these systems addressed the interests of a variety of participants, including the renter and the small holder as well as the large landholder. These participants took pragmatic views of how they wanted these systems to operate and the mechanisms they employed to reach the ends they envisioned. As the principles adumbrated above suggest, villagers strove to act on a sense of "fair share" rather than absolute equity, even in "equal distribution" types of *warichi*. The opportunity for gain from personal investment was usually still present under these systems and was accommodated differently as times changed, even into the twentieth century. Joint ownership systems and operations changed over time, reflecting repeated reevaluation of practice, with outcomes of any given iteration the result of a combination of negotiation, persuasion, and pressure. Joint ownership was compatible with economic growth and diversification during the seventeenth to late nineteenth centuries, a history that suggests that some kind of intermediate form of ownership between public and private may provide opportunities for effective resource management even today.

## Coda

Over the past century, Japan has changed tremendously but not always quite as much as we suppose and not everywhere in a uniform manner or at a lockstep pace. *Warichi*, the allocation of access to arable land by lot and other mechanisms, is now almost completely erased from the rural scene. Yet the allocation of access to individual plots by lot, a parallel to the most prominent of *warichi* allocation strategies, played an important part in the rationalization of agricultural fields that has been a major focus of late-twentieth-century agricultural policy in Japan. As those who traveled Japan's countryside during the field rationalization process know, to create fields of a consistent size and shape suitable for efficient, mechanized rice agriculture, old paddy ridges and paddies were bulldozed, new ridges erected, and paddy beds conditioned. This process destroyed the boundaries of individual fields and created a new geography of agriculture. Families who wished to continue farming drew lots for the new fields in proportion to their prior holdings.

In other instances, during the 1970s, the need to construct new dikes to control flooding often led the state to take over fields where residents still practiced *warichi*. It removed lands from agricultural use and constructed larger, cement dikes. In a number of instances these lands became general recreational areas between dike and stream.

Before construction, local officials approached neighborhood leaders to discuss how to deal with the state's imminent exercise of eminent domain. Residents did not always readily accede to the confiscation of their lands, but the state eventually won. In many instances, compensation received from the state was simply divided among the shareholding families, reflecting the predominant economic attitude of high-growth Japan.

The shareholders of Seiriki in Kumayama, too, met to discuss how to deal with the funds they would receive in compensation for the loss of their lands. Some residents made a strong case to simply divide the proceeds for families to use as they pleased. Their position did not carry the day. Invoking a sense of community values, one individual emulated Nagaoka domain's famous "education" counselor, Kobayashi Torasaburo, and persuaded his neighbors that since the jointly owned land had been transmitted for generations for the benefit of the community as a whole, the monies received in compensation for the land should go into a fund to support the education of future generations of the neighborhood rather than being divided among hamlet residents.[28] In this community, the spirit that modern Japanese saw in their forbears' joint ownership practice continues to benefit its residents.

# Appendix A:
## Sources for Redistribution Interval Data and Coordinate Data, Echigo Villages

This table includes data for all villages for which regular redistribution interval data were located, including several villages for which I was not able to locate latitude and longitude coordinates and that are therefore not displayed on my maps in Chapter 6. In addition, where a source indicated intervals for a larger area (county), I have included those data even though they were not used in my maps.

| Edo Era Village Name | Alternate Coordinate Map Source (all 1:50,000 Topographic Series) | Standard Interval | Comments | Alternate Interval Source |
|---|---|---|---|---|
| Akadzuka | | 7 | same interval for the Sakata hamlet (ōaza) | |
| Aoki | Taishō Nagaoka | 4 | | |
| Aratokawasawa | | 10 | | Yoshikawa Chōshi Hensan Iinkai 1, 707 |
| Arinobu | | 4 | | Teradomari Chōshi Hensan Iinkai, Tsūshi 1, 358 |
| Asahi | | 7 | | |
| Atsuta | | 3 | same interval for the Meishō hamlet (ōaza) | |
| Chūjō | | 7 | | Yoshikawa Chōshi Hensan Iinkai 1, 707 |
| Doai | Taishō Nagaoka | 10 | paddy, dry field, residential garden lands | |
| Doguchi | | 10–15 | | |
| Fukui | | 5 | but Niigata-ken Nōgyō, p. 140: Fukui village exchanged paddy and dry field once every 10 years according to discussions with old-timers; mountain forest lands once every 20 years | |
| Fukumichi | Taishō Nagaoka | 10 | | |

| Place | | Interval | Notes | Source |
|---|---|---|---|---|
| Funasaka | | 12 | | Tokamachi-shi Shi Hensanshitsu, *Tsūshi*, 3, 16 |
| Futsukamachi | | 6 yrs paddy; 10 yrs dry | | Tochio-shi Shi Hensan Iinkai, *jō*, 441, chart 45 |
| Ganjima | Taishō Nagaoka | 10 | same interval for the Kawabukuro hamlet (*ōaza*) | |
| Gōmoto | | 5 | | Teradomari Chōshi Hensan Iinkai, *Tsūshi*, 1, 358 |
| Gosengoku | Taishō Sanjō | 10 | Yata Gobunnoichi hamlet (*ōaza*) | |
| Hachibuse | Taishō Nagaoka | 4 | | |
| Hasugata | Taishō Nagaoka | 5 | redrawing of lots 5 years, full reassessment/drawing every 10 years | |
| Higashi Nakanomata | | 2 | for reclaimed land; interval for base fields unclear | |
| | | | base fields 3 years, reclaimed 2 | Tochio-shi Shi Hensan Iinkai, *jō*, 441, chart 45 |
| Higashi Yoshio | | 10–15 | | |
| Hirajima | | 20 | | |
| Hirano | | 5 | | |
| Hitodzura | | 10 | | Tochio-shi Shi Hensan Iinkai, *jō*, 441, chart 45 |
| Idzuka | | 5 | | |
| Iwade | | 10 | | |

| Edo Era Village Name | Alternate Coordinate Map Source (all 1:50,000 Topographic Series) | Standard Interval | Comments | Alternate Interval Source |
| --- | --- | --- | --- | --- |
| Kagi | Taishō Nagaoka | 4 | | |
| Kaji | | 5 | 8 for reclaimed land | Yoshikawa Chōshi Hensan Iinkai 1, 707 |
| Kakumi | | 5 | | Maki-machi Shi Hensan Iinkai, *Tsūshi, jō*, 450–451; also Mitsuke-shi Shi Hensan Iinkai, 628 |
| Kakumi | | | variation by vegetation and use; permanent; expansion of permanent division over time for mountains | |
| Kamagajima | Taishō Nagaoka | 8 | paddy | |
| | | 10 | dry field | |
| | | 10 | residential garden | |
| Kamegai | | 30 | excludes residential, hemp fields, drying racks, seedbeds | Niigata-shi Shi Hensan Iinkai, *Shiryō* 4, 340–341, document 109; Kurosaki Chōshi Hensan Iinkai, *Tsūshi*, 219–220 |
| Kamihimizo | | 13 | | Nakazato Sonshi Senmon Iinkai 1, 733 |
| Karasugajima | | 4 | | Tochio-shi Shi Hensan Iinkai, *jō*, 441, chart 45 |
| Karui | Taishō Sanjō | 3 | | |

| | | | | |
|---|---|---|---|---|
| Kawaji | | 4 | | Tokamachi-shi Shi Hensanshitsu, *Tsūshi* 3, 15 |
| Kiba | | 10 | | Kurosaki-chō Shi Hensan Iinkai, *Tsūshi*, 219–220 |
| Kijima | | 8 | hamlets (*ōaza*) Manzenji and Shimokiri, same | |
| Kita Daishi | | 10 | | Yoshikawa Chōshi Hensan Iinkai 1, 707 |
| Kitadani | | 10–15 | | |
| Kiyamazawa | | 2 | | Tochio-shi Shi Hensan Iinkai, *jō*, 441, chart 45 |
| Kodomari | | 8 | | |
| Koguriyama | | 8 | | |
| Koshin | | 15 | | |
| Koshōji | Taishō Nagaoka | 10 | | |
| Machida | Taishō Nagaoka | 4 | | |
| Machikarui | | 5 | | Teradomari Chōshi Hensan Iinkai, *Tsūshi* 1, 358 |
| Magiyama | | 20 | | |
| Magoshi | | 10 | redrawing of lots; full evaluation and redistribution 30 years | Tsukamoto 1992b, 43 |
| Makiyama | | 5 | Originally 10 years, changed later, date unclear | |

| Edo Era Village Name | Alternate Coordinate Map Source (all 1:50,000 Topographic Series) | Standard Interval | Comments | Alternate Interval Source |
|---|---|---|---|---|
| Masuzawa | | 10–15 | | |
| Matomegusa | | 4 | hamlet (*ōaza*) Niinaga, same | |
| Matsuda | | 4 | | Teradomari Chōshi Hensan Iinkai, *Tsūshi* 1, 358 |
| Matsunoo | | 5 | Matsuyama Shinden, Ohara shinden, same | |
| Mikata | | 8 | | |
| Minakuchi | | 10–15 | | |
| Minamishima | Taishō Nagaoka | 4 | hamlet (*ōaza*) Jōhana, same | |
| Nadachi ōmachi | | | | |
| Nagakura | Taishō Nagaoka | 4 | | |
| Nakajima | | 20 | | |
| Nakajo | | 10 | Tsumefu and Shinbō hamlets (*ōaza*), same | |
| Nakamura | | 2 | | Tochio-shi Shi Hensan Iinkai, *jō*, 441, chart 45 |
| Nakazawa | | 20 | | |
| Natsuto | | 2 | | Teradomari Chōshi Hensan Iinkai, *Tsūshi* 1, 358 |

| | | | | |
|---|---|---|---|---|
| Nigoro | | 3 | | Tochio-shi Shi Hensan Iinkai, *jō*, 441, chart 45 |
| Nirebara | | 2 | seems to be the most common interval documented, but longer intervals present | Tochio-shi Shi Hensan Iinkai, *jō*, 454–455 |
| Nishi Nakanomata | | 3 | | Tochio-shi Shi Hensan Iinkai, *jō*, 441, chart 45 |
| Nishinoshima | | 8 | | |
| Nishi Yachi | | 10–15 | | |
| Nishi Yoshio | | 10–15 | | |
| Nonakasai | | 10 | | Tsukamoto 1992b, 45 |
| Ochiai | | 10 | | |
| Ogino | Taishō Nagaoka | 5 | | |
| Ono | | 4 | | Tochio-shi Shi Hensan Iinkai, *jō*, 441, chart 45 |
| Ōbuchi | | 10–15 | | |
| Ōji | | 20 | | Teradomari-chō Shi Hensan Iinkai, *Tsūshi* 1, 358 |
| Ōmachi | Taishō Nagaoka | 4 | | |
| Ōmagari | | 12 | | |
| Ōwada | | 5 | reclaimed = 6 | Teradomari Chōshi Hensan Iinkai, *Tsūshi* 1, 358 |
| Ōzeki | | 5 | | |

| Edo Era Village Name | Alternate Coordinate Map Source (all 1:50,000 Topographic Series) | Standard Interval | Comments | Alternate Interval Source |
|---|---|---|---|---|
| Raikōji | | 10 | hamlets (*ōaza*) Kitano and Nishino, same | |
| Rokumanbu | | | permanently converted to private holding, no redistribution | |
| Sannōfuchi | | 10 | | Tsubame-shi Shi Hensan Iinkai, *Tsūshi*, 292 |
| Sekiya | | 10 | | Niigata-shi Shi Hensan Iinkai, *Shiryō* 4, 349–357, document 117 |
| Shindōri | | 10 | | Niigata-shi Shi Hensan Iinkai, *Shiryō* 4, 342–343, document 111 |
| Somagi | Taishō Sanjō or Taishō Yahiko | 5 | | |
| Sugebatake | | 5 | | Tochio-shi Shi Hensan Iinkai, *jō*, 441, chart 45 |
| Sugisawa | | 3 | | Mitsuke-shi Shi Hensan Iinkai, *jō*, 627 |
| Tai | Taishō Nagaoka | 2 | | |
| Tajiri | | 10 | paddy and dry field | |
| | | 20 | mountain woodlands | |
| | | 4 | | |

| Village | Taishō designation | Number | Notes | Source |
|---|---|---|---|---|
| Takabatake | | 4 | | |
| Takano | Taishō Nagaoka | 10 | implicit in dispute documents | Nagaoka-shi Shi Hensan Iinkai, *Tsūshi* 1, 497 |
| Takauchi | | 10 | Source indicates up to 15 | Teradomari Chōshi Hensan Iinkai, *Tsūshi* 1, 358 |
| Takenao | | 8 | 16 for one small reclaimed land place | Yoshikawa Chōshi Hensan Iinkai 1, 707 |
| Takenomachi | | 8 | | |
| Tanaka | | 10 | | Yoshikawa Chōshi Hensan Iinkai 1, 707 |
| Tanokuchi | | 2 | | Tochio-shi Shi Hensan Iinkai, *jō*, 441 chart 45 |
| Tochibori | | 4 | this is longest, changes in some cases to 2 years or every year; some indication on this page that every year rotations or *kujikae* led to free rider issues; 453 indicates several villages that stopped both *kujikae* and *warichi* in mid-18th century | Tochio-shi Shi Hensan Iinkai, *jō*, 452 |
| Toshitomo | | 2 | *Hatake* is permanently held | Teradomari Chōshi Hensan Iinkai, *Tsūshi* 1, 358 |
| Tsubono | | 13 | | Yoshikawa Chōshi Hensan Iinkai 1, 707 |
| Tsuboyama | | | | |
| Uchino | | 3 | | |
| Urase | Taishō Nagaoka | 5 | | |
| Wakigawa Shinden | Taishō Sanjō | 1 | | |

| Edo Era Village Name | Alternate Coordinate Map Source (all 1:50,000 Topographic Series) | Standard Interval | Comments | Alternate Interval Source |
|---|---|---|---|---|
| Wakinomachi | | 10 | every 10 years full redistribution and additional reclaimed land; exclusions: mountain woodlands, crop-drying racks, seedbeds, and mountain dry fields | |
| Watabe | | 10 | seedbeds, 20 | |
| Yamabe | | 20 | | |
| Yamaya | | 25 | | Tokamachi-shi Shi Hensanshitsu, *Tsūshi* 3, 16 |
| Yamazawa | Taishō Nagaoka | 4 | | |
| Yokobatake | | 10–15 | | |
| Yokoyama | | | permanently converted to private holding at some point | |
| Nishi Kanbara-gun | | 10 | paddy, dry field, residential garden lands | |
| Nadachi General | | 5 | paddy | |
| | | 10 | dry field | |
| | | 20 | residential garden | |
| | | 10 | mountain woodlands | |

*Sources:* Interval data source: Niigata-ken Naimubu 1929, unless otherwise noted; if not noted otherwise in column 2, latitude/longitude coordinate source is Kokudo Chiriin 1982, 1:25,000 Map Series Place-Name Database.

# Appendix B:
## Hikikuji Usage, Supplementary Tables

**Table 1**  Cultivator *Hikikuji* Exemption Usage by *Wari* and Land Type, 1781

| | PADDY | | | DRY | | | |
|---|---|---|---|---|---|---|---|
| VILLAGER | SHIMONISHI 100-*BU WRI* | URANAKA 2 *WRI* | GOTANDA 2 *WRI* | KAMI(?) NISHI 30-*BU WRI* | KAMI(?) NISHI 2 *WRI* | ZANBU 30-*BU WRI* | TOTAL |
| Original holder not designated; presumably document author. Current holders also not always clear. | 1 | 1 | 3 | 1 | 2 | 2 | 10 |

*Source:* Satō-ke monjo, document 8014.

**Table 2**  Cultivator *Hikikuji* Exemptions Usage by Wari and Land Type, 1809

| Villager | Waseda 60 bu | Shimonishi 25 bu | Himetsuruta 200 bu | Uranaka 1 | Uranaka 2 | Gotanda 1 | Gotanda 2 | Sawada 70 bu | Zanbu 50 bu 2 | Total |
|---|---|---|---|---|---|---|---|---|---|---|
| | | | PADDY | | | | | | | |
| Zenbei | 1 | 1 | | | | | | | | 2 |
| Zenbei, Sanueimon | | | 1 | | | | 1 | | | 2 |
| Sakuzaeimon | | | | 1 | | | | | | 1 |
| Kitarō | | | | | 2 | 2 | 4 | | | 10 |
| Shōzaeimon, Yohachi, Kitarō | | | | | | 1 | | | | 1 |
| Sakuzaeimon, Fujizaeimon, Kitarō | | | | | | 1 | | | | 1 |
| Yasuzaeimon | | | | | | | | | 1 | 1 |
| Sahachi | | | | | | | | | | 0 |
| Senzō, ? | | | | | | | | | | 0 |
| Yahachi, Shōzaeimon | | | | | | | | | | 0 |
| Yozaeimon, Shōkichi | | | | | | | | | | 0 |
| Denbei | | | | | | | | | | 0 |
| Takahara | | | | | | | | | | 0 |
| Yoshiueimon, Jūbei | | | | | | | | 1 | | 1 |
| Unclear | | | | | | | | 2 | | 2 |
| Total | 1 | 1 | 1 | 1 | 2 | 6 | 5 | 3 | 1 | 21 |

*Names of exemption holders in these columns have been crossed out in the document; nontheless, I

*Source:* Satō-ke monjo, document 8374.

| Villager | _Asabatake 30 bu 1_ | _Asabatake 30 bu 2_ | _Shinden-batake 50 bu_ | _Takehara-hatake 50 bu 1*_ | _Takehara-hatake 50 bu 2*_ | _Takehara-hatake 50 bu 3*_ | _Takehara-hatake 100 bu 2_ | _Takehara-hatake 50 bu 1_ | _Nokori-batake 30 bu_ | Total |
|---|---|---|---|---|---|---|---|---|---|---|
| Zenbei | 3 | 1 | | | | | 1 | 2 | | 7 |
| Zenbei, Sanueimon | | | | | | | | | | 0 |
| Sakuzaeimon | | | | | | | | | | 0 |
| Kitarō | 2 | 2 | | (2) | | | | 2 | | 6 (8) |
| Shōzaeimon, Yohachi, Kitarō | | | | | | | | | | 0 |
| Sakuzaeimon, Fujizaeimon, Kitarō | | | | | | | | | | 0 |
| Yasuzaeimon | 2 | | | | | | | | | 2 |
| Sahachi | 1 | | | | | | | | | 1 |
| Senzō, ? | 1 | | | | | | | | | 1 |
| Yahachi, Shōzaeimon | 1 | 1 | | | | | | | | 2 |
| Yozaeimon, Shōkichi | | 1 | | | | | | | | 1 |
| Denbei | | | 2 | | | | | | | 2 |
| Takahara | | | 1 | | | | | | | 1 |
| Yoshiueimon, Jūbei | | | | | | | | | | 0 |
| Unclear | | | | | (2) | (3) | 3 | | 2 | 5 (10) |
| Total | 10 | 5 | 3 | (2) | (2) | (3) | 4 | 4 | 2 | 28 (35) |

have treated them as representing actually exercised exemptions.

**Table 3**  Cultivator *Hikikuji* Exemptions Usage by *Wari* and Land Type,

| Villager | Shimonishi 25 bu | Shimonishi 25 bu | Waseda 60 bu | Uranaka 1 | Uranaka 2 | Gotanda 1 | Gotanda 2 | Zanbu 50 bu 2 | Total |
|---|---|---|---|---|---|---|---|---|---|
| | Paddy | | | | | | | | |
| Hachibei | 1 | 1 | | | 2 | 4 | 4 | 4 | 16 |
| Zenbei | | | 1 | 2 | | | | | 3 |
| Denbei | | | | | | | | | 0 |
| Total | 1 | 1 | 1 | 2 | 2 | 4 | 4 | 4 | 19 |

*Source:* Satō-ke monjo, document 8012.

1818 *Kujikae chō*

| Villager | DRY FIELD | | | | Total |
| | Takehara-batake 100 bu 1 | Takehara-batake 100 bu 2 | Takehara-batake 100 bu 3 | Shinden 60 bu 3 | |
|---|---|---|---|---|---|
| Hachibei | 2 | | | 2 | 4 |
| Zenbei | 1 | 1 | 2 | | 4 |
| Denbei | | | | 2 | 2 |
| Total | 3 | 1 | 2 | 4 | 10 |

**Table 4** Cultivator *Hikikuji* Exemptions Usage by *Wari* and

| | PADDY | | | | | |
|---|---|---|---|---|---|---|
| VILLAGER | SHIMONISHI 100 BU | URANAKA 1 | GOTANDA 1 | GOTANDA 2 | ZANBU 50 BU 2 | TOTAL |
| Hachibei | 1 | | 4 | 4 | 4 | 13 |
| Zenbei | | 2 | | | | 2 |
| Denbei | | | | | | 0 |
| Denbei, Yoshiueimon | | | | | | 0 |
| Eitarō | | | | | | 0 |
| Sahachi | | | | | | 0 |
| Hyōeimon, Tadazaeimon | | | | | | 0 |
| Shōzaeimon, Yahachi | | | | | | 0 |
| Kishichi, Yūzaeimon | | | | | | 0 |
| Total | 1 | 2 | 4 | 4 | 4 | 15 |

*Source:* Satō-ke monjo, document 8375.

Land Type, 1818 *Gechō*

| Villager | Dry Field | | | | | | | | | Total |
| | Takehara-batake 100 bu 1 | Takehara-batake 100 bu 2 | Takehara-batake 50 bu 1 | Hatake 60 bu | Nagatoro 45 bu | Kawaharanoko-ri-batake 40 bu | Asabatake 30 bu 1 | Asabatake 30 bu 2 | Nokori-batake 30 bu | |
| --- | --- | --- | --- | --- | --- | --- | --- | --- | --- | --- |
| Hachibei | 2 | | | 2 | | | 2 | 2 | 2 | 10 |
| Zenbei | 1 | 1 | 2 | | 1 | 1 | 3 | 1 | | 10 |
| Denbei | | | | 2 | | | | | | 2 |
| Denbei, Yoshiueimon | | 1 | | | | | | | | 1 |
| Eitarō | | | | | | | 1 | | | 1 |
| Sahachi | | | | | | | 1 | | | 1 |
| Hyōeimon, Tadazaeimon | | | | | | | 1 | | | 1 |
| Shōzaeimon, Yahachi | | | | | | | 1 | 1 | | 2 |
| Kishichi, Yūeimon | | | | | | | | 1 | | 1 |
| Total | 3 | 2 | 2 | 4 | 1 | 1 | 9 | 5 | 2 | 29 |

**Table 5**  Cultivator Paddy *Hikikuji* Exemptions Usage by *Wari* and Land Type, 1842

| VILLAGER | WASEDA 60 BU | URANAKA 1 | URANAKA 2 | GOTANDA 1 | GOTANDA 2 | TOTAL |
|---|---|---|---|---|---|---|
| Yūshichi | 1 | 1 | | | | 2 |
| (Yūshichi) Sōzaeimon | | 1 | | | | 1 |
| (Teisuke's man) Gonzaeimon | | | 2 | | | 2 |
| Takahara, Chūzaeimon | | | | 1 | | 1 |
| Yonekichi, Eijirō | | | | 1 | | 1 |
| (Takahara's man) Ninbei | | | | 1 | | 1 |
| Jūbei, Yoshibei | | | | 1 | 1 | 2 |
| Takahara, Sakuzaeimon | | | | | 1 | 1 |
| Total | 1 | 2 | 2 | 4 | 2 | 11 |

*Source:* Satō-ke monjo, document 8379.

**Table 6**  *Hikikuji* Exemptions Usage by *Wari*, 1 Paddy, Date Unclear

| VILLAGER | URANAKA 1 | GOTANDA 1 | GOTANDA 2 | TOTAL |
|---|---|---|---|---|
| Unclear | 2 | 3 | 3 | 8 |
| Total | 2 | 3 | 3 | 8 |

*Source:* Satō-ke monjo, document 8017.

**Table 7**  *Hikikuji* Exemptions Usage by *Wari*, 2 Paddy, Date Unclear

| VILLAGER | WASEDA 60 BU | URANAKA 1 | URANAKA 2 | GOTANDA 1 | GOTANDA 2 | TOTAL |
|---|---|---|---|---|---|---|
| Yūshichi | 1 | | | | | 1 |
| Jūbei, Yoshibei | | | | 1 | 1 | 2 |
| Chūzaeimon | | 1 | | 1 | 1 | 3 |
| Eijirō, Yonekichi | | | | 1 | 1 | 2 |
| Teisuke | | | 2 | 1 | 1 | 4 |
| Total | 1 | 1 | 2 | 4 | 4 | 12 |

*Source:* Satō-ke monjo, document 8264.

# *Notes*

**Preface**

1. Brown 1993, chapter 4.

**Chapter 1: Introduction**

1. Similar approaches were sometimes also taken to marine resources in Japan's coastal areas, rivers, and lakes.

2. Brown (various); Lieban 1956; McKean 1985 and 1989; and Lewis 1985 represent exceptions.

3. While excessive moisture in a given soil structure and vegetation cover is often associated with landslides, other factors, notably earthquakes, are also major stimuli.

4. European open field systems embodied similar portfolio effects. See McCloskey 1975a and 1975b.

5. Pallot 1999.

6. Ooms 1996 does not examine agriculture. Other works deal with rural society in some measure, notably a clutch of publications on "peasant rebellions" in the late 1980s and early 1990s. They explore rural crises that provide some background on villagers' lives.

7. Stiglitz 1974.

8. Libecap 1989 brings together a number of his studies.

9. Dahlman 1980, especially p. 167, offers a property rights analysis of the open field system and enclosure movements in England. He notes that the

informal and local enclosure movements were so widespread that the formal acts of parliament were arguably just a ratification of processes that already had largely eliminated open fields.

10. Geographer Judith Pallot (1999) presents a detailed analysis of several case studies.

11. See, for example, Purnell 1999.

12. Solnick 1998 discusses the disruptive outcomes of privatization and property rights transformation in post-Soviet Russia.

13. Reform of China's land rights laws and institutions continues. See Schwarzwalder 2000 and Schwarzwalder et al. 2002.

14. Hardin 1968.

15. Later, Hardin (1991) retreated from this position, acknowledging the commons could be successful.

16. Brubaker 1975, 147.

17. A selection of major thinkers is conveniently available in Ackerman 1975. It is common to think of property rights as residing entirely within the grasp of their legal owner(s) although there are limited circumstances in which a property right can be split, such as subsurface rights to natural gas.

In contrast, Barzel 1989, written by an economist, posits that what we call a property right is really a bundle of many related rights, only some of which are recognized and deemed to reside in an owner. Property rights are captured only when they are valuable enough to cover enforcement costs.

18. For one example by a number of specialists, including historians, see Richards 2002. The volume concludes that property rights alone do not determine resource maintenance outcomes. Astute social scientists, including Nobel laureate (economics) Elinor Ostrom, have devoted considerable effort to exploring the circumstances under which common pool resource management has been viable. See Ostrom 1990.

19. Miller 1992.

20. Brown and Taniguchi 2000; Brown 2003a, 2003b, and 2002a.

21. Smith 1959; Pratt 1999.

22. Smith 1959; Hanley and Yamamura 1977.

23. For Europe, see North and Thomas 1973.

24. Hayami 2004, 9–10.

25. Ibid., 28. For examples of this research, see the essays in Hayami, Saito, and Toby 2004. Other articles appear in the eight volumes of Hayami et al. 1988–1990 and related earlier publications of the Quantitative Economic History Group.

26. See his review of family structural adaptability in Saitō 2000, esp. pp. 29–30, on the decline of long-term service contracts. See also Takagi 2000.

27. Care must be taken in using Western legal concepts to describe concepts of landholding in pre-Meiji Japan. Today, eminent domain is generally used in societies on a limited basis and involves the transfer of usufruct from

the hands of private individuals or groups to a public agency to meet some specific public good. Our Western conception of eminent domain implies a clear legal distinction between "public" and "private" ownership. It also implies the right of the "private" owner to receive compensation. Daimyo took lands from local village control and reserved them for domain use but, as in eighteenth-century Great Britain, did not provide compensation. Perhaps the most widespread example of this process is the establishment of domain forests for military practice and timber supply. At lower levels of administration, districts or villages might likewise require access to a villager's land to build new irrigation works. Again, the object of public control was limited in scope, and the public function was clear. (*Warichi* and related systems did not take usufruct away from a household but reassigned it.) On domain forests, see Totman 1984 and 1989. For Tokugawa irrigation systems see Kelly 1982.

28. Ishii 1966, 150–202. Ishii made no examination of the precision and replicability of actual land survey techniques.

29. Yamamura 1981, 283. In contrast to Yamamura's emphasis on the economic incentives, Japanese scholars, such as Araki Moriaki, see centrally initiated surveys as a coercive device, redistributing property rights in order to squeeze more out of the peasantry. Araki 1959, 181–186. Similarly, see Miyagawa 1959 and Sasaki 1964. I discuss this literature on the Taikō *kenchi* in Brown 1993, 10–24.

30. Hall 1981, 339–349. Other stimuli to efficiency are also discussed.

31. Ibid., 212, 213, 194.

32. Broadbent 1966; Blackford 1988.

33. Francks 2006.

34. Totman 1989 and 1984; Walker 2006; Howell 1995.

35. Kelly 1982; Walthall 1986; White 1995; and Ooms 1996.

36. The classic literature in the field has already been reviewed in Smith 1959 and Hanley and Yamamura 1977, so I abbreviate discussion here.

37. These reforms aimed to increase the authority of the emperor through adoption of Chinese patterns of administration, including the system of taxation.

38. Margaret McKean has written extensively on these practices in the Mt. Fuji region. See McKean 1989; see also Lewis 1985.

39. In addition to McKean's work, one of the best-known studies is Smith 1959. See also Smith 1968; Befu 1968; and McKean 1985. For village and district control over irrigation systems, see Kelly 1982. For forests, see Totman 1984 and 1989.

40. Richard Lieban's 1956 doctoral thesis is the only exception of which I am aware.

41. Aono 1982.

42. See North 1990; Ostrom 1990; and the entire Cambridge University Press series the Political Economy of Institutions and Decisions.

43. Kawata 1912 specifically discusses the link between ownership rights and emotional attachment to land but does not consider *warichi*.

44. A small selection of these documents are cited in the chapters that follow, but a perusal of indexes to the archive collections cited below reveals hundreds, if not thousands, of such documents for Echigo alone.

## Chapter 2: Origins and Geopolitical Contexts

1. Other than my own work and that of anthropologist Richard Lieban, there is no English-language scholarship on *warichi*.

2. See Furushima 1939.

3. Oishi 1969, *jō*, 14, 123.

4. Nakada 1904. Uchida 1921 also notes the association between land reclamation and joint land tenures.

5. Brown 1988, 1993.

6. The "moral economy" approach in Asian studies is presented in Scott 1976; Popkin 1979 presents the alternative argument.

7. Aono 1982, especially pps. 1–2 *(maegaki)*, 17–19, 35–38, outline this interpretation. Each chapter examines this hypothesis in a case study.

8. Ibid., 17–19.

9. Aono 1982; Brown 1987a.

10. Kokudo Chiriin 1982.

11. Trewartha 1965, 22, notes that there are differences between Japanese and non-Japanese classifications of tectonic zones; Trewartha stresses four zones, the Japanese more.

12. E.g, Niigata-shi Shi Hensan Iinkai 1997, *Tsūshi hen 2*, 369–373.

13. Bird 1984, 192–197.

14. Brown 2008a.

15. Brown 1993; Ravina 1999; Roberts 1998.

16. Hauser 1985.

17. Brown 1988 demonstrates that in some significant instances the shogunal administration was the follower.

18. Brown 1993.

19. Larger domains were generally located a considerable distance from the shogun's castle town at Edo (modern Tokyo), further restricting the possibility of effective shogunal oversight.

20. Brown 1988.

21. A few domains inserted samurai into villages, Satsuma and Tosa, for example, but this did not necessarily mean that samurai oversaw village affairs because samurai typically were not stable village residents, sometimes being rotated regularly (as in the case of Satsuma), at other times successfully fighting domain rustification policies and moving back to the castle town (Hirosaki domain, northern Japan).

22. Ravina 1999; Jansen 1968.

23. Although initially selected by daimyo administrators, such positions were inherited through multiple generations of the same family, like the office of village headman.

24. In some domains, some county-level administrators were, in fact, commoners. For Kaga's use of commoners, see Brown 1993, 124–131, 133–141, 172–178, for example.

25. The variation in shogunal practice is well known. For Tōdō domain, see Fukaya 1969.

26. Several recent survey treatments by prominent scholars illustrate this point. James McClain makes reference to daimyo autonomy but within a delimited sphere and saves his most detailed descriptions of administration for shogunal structures. Conrad Totman specifically devotes space to discussing developments in several domains, describing each briefly, but also devotes his greatest energy to descriptions of the shogun. McClain 2001, especially pp. 24–30; Totman 1993.

27. Jansen 2000, chapter 2.

28. Brown 1993 along with sections of Ravina 1999 provide additional evidence on the importance of local, rather than shogunal, initiative.

29. Brown 1993, 110.

30. From the early seventeenth century, Sado was administered directly by the shogun in order to exploit its rich silver and gold mines ( Japan's richest).

31. Niigata-shi Shi Hensan Iinikai 1989–1998, *Shiryō hen* 4, Niigata, Japan, 1311–1312; see also ibid., *Tsūshi* 2, "Preface."

32. See Pratt 1999.

33. Niigata-ken Shi Hensanshitsu. 1990, 173–175. The classic English-language study is Smith 1966; see also Smith 1968; see Bray 1986 for a more extended essay on the characteristics of intensified rice agriculture.

An outline of the changes in agricultural practices described in Japanese agriculturalists' writings is provided by Furushima 1961 and Sugimoto 1976. For detailed treatment of developments in the Kaga domain and Echigo regions, see Furushima 1975 and Arashi 1975. Arashi's work seldom mentions changes in agriculture specifically in Satsuma, but his extensive treatment of the role of new seed varieties in rice cultivation in southwest Japan demonstrates throughout that developments elsewhere in Kyushu also appeared in Satsuma, a joint ownership region.

34. Brown, 1987b.

35. Fukushima 1968, 253–255, presents a brief summary of this issue. Arimoto 1967 presents selected statistical data from villages and early Meiji prefectures that composed the mainland part of later Niigata Prefecture (Niigata, Kashiwazaki), which reinforce the impression that the change was particularly significant for Niigata; see pp. 273, 410–411. Kondō 1967, 115, presents regionally aggregated data.

36. Niigata-ken Shi Hensanshitsu 1990, 175–176.

37. The *sen chōbu jinushi*; 1,000 *chō* (or *chōbu*) is the equivalent of about 2,450 acres or just over 9,900 square meters, just under ten hectares.

38. Nihon Kasen Kyōkai 2005, 84.

39. White 1995, 72–73.

40. Suetsugi 2005, 29.

41. Small field size also plays a significant role in paddy agriculture: it reflects the need to keep a flat pan so that a field's water level will be uniform—a requirement that is more readily filled on small fields than large.

42. Waters 1983; Kelly 1982.

43. Smith 1959 is the classic study.

## Chapter 3: Data and Methodologies

1. For illustrative examples of GIS use in history-related research and teaching, see Knowles 2002, 2005, and 2008; Brown 2008b.

2. Furushima 1939; Aono 1982.

3. My inquiries about *warichi* led Ishida to include brief discussions of these practices in Kumayama Chōshi Hensan Iinkai 1993.

4. In societies like England, royal courts provided a mechanism for preserving records of disputes that reveal a great deal about lower social orders. For an English example, see Hanawalt 2007.

5. Wigmore 1967–1984; Henderson 1975.

6. Other than one or two illustrative examples, documents that contain statistical data are seldom transcribed and published in local histories.

7. I am not entirely alone in this approach: some Japanese scholars in the 1950s and 1960s also took advantage of such opportunities, as did occasional newspaper reporters.

8. Other maps were created from print and outline maps that I modified using Adobe's Photoshop and Acrobat programs.

9. Siebert 2000, 207–224.

10. There are some exceptions, such as the well-known dispersed settlement pattern of villages in Tonami-gun (county) in modern Toyama Prefecture.

11. Yoshikawa Chōshi Hensan Iinkai 1993–1996.

12. Brown 2005.

13. This is not simply a matter of scale. Zooming in shows a divergence of only 3 to 5 meters between the sets of data. See the illustration at the following URL: http://people.cohums.ohio-state.edu/brown113/. Follow the GPS vs. Japan Gov. Data Comparison close-up link at page bottom.

14. Three Japanese colleagues are currently working on versions of this issue. Murayama Yuji of Tsukuba University has been developing a GIS database of changing city, town, and village boundaries going back to the mid-Meiji era. When complete, this data will encompass something under 16,000

low-level administrative units. Mizoguchi Tsunetoshi, Nagoya University, is engaged in digitizing maps made of early Meiji villages as part of the Meiji Land Tax Reforms. While these maps were not the product of modern cartographic surveys, boundaries can be adjusted in a GIS environment through a process called rubber sheeting, which uses algorithms to approximate locations using known latitude and longitude of sample points. Kawaguchi Hiroshi of Tezukayama University takes a different approach. Noting that the total number of late Tokugawa villages listed in national compilations exceeds 63,500, far more than the number of maps that exist for the Meiji Land Tax Reforms, he is following the process for locating hamlet centers that I have outlined here. Once he has location data for hamlet centers, he will employ standard algorithms to approximate boundaries of these villages.

15. Locations for a tiny handful of villages could not be determined by these methods, and data from these cases have been omitted from the sample analyzed below. Although some of the hamlet names appeared as customary neighborhood names on these maps, the residential center of each hamlet was not always designated. In such cases, rather than using the latitude and longitude for the residential hamlet center, I used a point at the center of the place-name label.

## Chapter 4: Varieties and Extent of Joint Landownership

1. Fukushima 1968, 51–54.

2. Barzel 1989 posits that there is not a property right (singular) but a bundle of many related rights among which we legally recognize only those for which the benefits exceed the costs of capturing them.

3. Aono 1982 examines more than fifteen cases.

4. See Brown 1987b, 115–155, for a detailed discussion of typical land survey and assessment methods.

5. This discussion is based on the author's interviews with Kumayama-machi residents in fall 1993 and spring 1994. See Kumayama Chōshi Hensan Iinkai 1993, *Ōaza hen*, 267–269.

6. This practice bears some similarity to that of rotating marine fishing rights in Alanya, Turkey. Like Seiriki, the area in which rights were exercised was a narrow strip of the resource, in this case marine bottom squeezed between the shore and a steep trench. Rotation occurred daily. See Berkes 1980, especially 73 and 77–78.

7. Tenancy developed in this region during the Meiji era and after, but documents from the periods of origin are limited and do not reveal the social structure of villages.

8. Because of the old Japanese system of counting someone as one year old at birth, documents describing this age range will read "between sixteen and sixty years of age."

Although the *kadowari* system bears some resemblance to the *jōri* system, there is no evidence to support a connection between the two patterns of land allocation. The system is often treated as one of Satsuma's "exports" to Okinawa; however, there are scholars who argue that the system predates the Shimazu dominance. See Umeki 1991. My limited investigations of *warichi* in Okinawa indicate that there were multiple varieties, a fact that argues against a top-down imposition. For example, in Kudakajima allocation took into consideration the number of minor children and women in the village, a clear departure from the *kadowari* model.

9. See Ono 1931, 72–75.

10. Matsushita 1984, 209–233, argues that in Satsuma there were no actual measurements associated with Hideyoshi's tallies of land value.

11. Ono 1931, 51–52.

12. Sakai 1968.

13. Ono 1931, 40–46, 56. Villager cultivation of the headman's lands suggests a medieval continuation of dependent labor relationships.

14. Ibid., 57–58.

15. Ibid., 22–23.

16. *Myōzu* are thus different from *myōshu*, a term commonly used in some parts of Japan to refer to large landholders.

17. Ono 1931, 57–58, 61–66. *Myōzu* generally held from 20 to 30 percent of the land.

18. Ibid., 59, 66–67.

19. In addition to those conducted by Hideyoshi's agents in 1594, investigations were implemented in all villages in 1604, 1633, 1659, and 1722 (ibid., 51–52).

20. Ibid., 53–55.

21. Ibid., 57. Calculation of this figure assumes that about 25 percent of the *kadodaka* was allocated to the *myōzu*.

22. Ibid., 30, 35.

23. Ibid., 35–36. Ono cites an undated document to support this contention, but provides no date for the start of the policy.

24. Ono 1931, 67–70, discusses the wide range of obligations the *kado* fulfilled.

25. Ibid., 22–23.

26. "Kawai roku," in Hanpō Kenkyū Kai 1966, 913; Tochinai 1936, 134–136.

27. Since the concern here is the movement from village to domain control, I limit discussion of the proof of early village use of *warichi*. Fuller discussion appears in Brown 1997, from which the Kaga section of this chapter is abstracted.

28. Heki 1970, 2:634.

29. For a fuller description of landed retainers' relationship to villages and daimyo, see Brown 1993.

30. Ibid., 196, 199–202, 204–215.

31. The documents from Ota village cited below provide two examples of *warichi* activity at this time. See also Tonami-shi Shi Hensan Iinkai 1965, 372–373. Document number 40, Wajima-shi Shi Hensan Senmon Iinkai 1971–1976, 1:473–474, is one example of the reporting of *teagedaka* during *warichi*.

32. For example, the 1679 request submitted by the farmers of Omachi. Hakui-shi Shi Hensan Iinkai 1975, *Kinsei hen*, 659.

33. Toyama-ken Shi Hensan Iinkai 1974–1985, *Shiryō hen, 3, Kinsei, jō*, 912.

34. Heki 1973, 603, entry for *denchiwari*.

35. Wakabayashi 1970–1972, *ge*, 147.

36. Heki 1973, 603. The 1838 regulations forbade the exclusion of such lands. Oda 1929, 491.

37. See the Tempo 9 (1838) regulations on *warichi* in Oda 1929, 490–492.

38. As noted above, a third category, *sōchi*, also crept into usage. Heki 1973, 603; Oda 1929, 491.

39. "Kawai roku," Hanpō Kenkyū Kai, 1966, 913.

40. In all but one Kaga domain county, paddy, residential lands, and high quality dry fields were all valued at 1.5 *koku* per *tan*. A number of dry fields would have been valued at half as much or less.

41. If someone had more residential lands than could be accommodated by his or her prorated share of this legal exemption, the remainder of that residential land could be treated as consuming part of the share of arable land. The *warichi* conducted in Awabara village in 1794 treated such lands as part of the allotment of average paddy. See the *sadamegaki*, Hakui-shi Shi Hensan Iinkai 1975, 665.

42. This exemption was called *kagebiki*.

43. Also called *san mono* or *sanyō mono*.

44. Domain surveys only lowered the assessed value of certain middling and inferior grades of dry field. All other lands were valued at 1.5 *koku* per *tan* except for Nōmi County at 1.7 *koku* per *tan*. Brown 1993, chapter 3.

45. The range of *wari* sizes was reduced slightly by an increase in the size of the 45-*bu wari* to 50 *bu*, creation of one new *wari* based on an increase in arable land reallocated (about 125 *bu*), and limited redefinition of the land in six of the 1812 *wari*.

46. Standard deviation in *wari* size had been reduced from 127.34 *bu* in 1812 to 79.86 *bu* in 1867.

47. These *wari* were replaced by four new ones that ranged in size from 300 *bu* to 320 *bu*. All four of the *wari* over 400 *bu* were reduced or eliminated.

48. Only three *wari* were smaller than 270 *bu*, and only one was 400 *bu*. Hakui-shi Shi Hensan Iinkai 1975, 675.

49. "Kawai roku," Hanpō Kenkyū Kai, 1966, 913.

50. Tochinai 1936, 136–142; "Kawai roku," Hanpō Kenkyū Kai, 1966, 913; Hakui-shi Shi Hensan Iinkai 1975, 671.

51. Wajima-shi Shi Hensan Senmon Iinkai 1971–1976, 1:55–57.

52. Hara 1971, 287.

53. Ideally, other divergent landownership practices such as those noted by Fukushima should be included, but these are beyond the scope of the present work.

54. Brown 2006.

55. Namely, Echigo, Echizen, Owari, Kaga, Noto, Etchu, Tosa, Ise, Tsushima, Hizen, Iyo, Iki, Hyuga, Satsuma, Mutsu, Hitachi, Tango, Yamato, Bungo, Chikuzen, Bizen, and Mino. See Furushima 1939, 134–162.

56. Aono 1982.

57. Further investigation of reallocation schemes in the advanced areas of Japan is needed to determine the extent to which these practices were confined to the more isolated parts of these regions.

58. Brown 1993, chapter 3, provides evidence for diverse methods of calculating *kokudaka*.

59. Kanai 1985.

60. It is best to think of the data in these different sources as reflecting independent estimates, not consistently compiled data that reflect changes in domain economic growth. Changes in domain value came from multiple sources, including accounting changes, land reclamation, and changes in the area a domain governed.

61. Aono 1982, 112.

62. Aono 1982 presents numerous examples; literature cited by Furushima 1939 presents others.

## Chapter 5: Lay of the Land: *Warichi* Practice in Iwade Village

1. Aono 1977, 101.

2. Funahashi 1995, 201–260; Takazawa 1979.

3. Niigata-ken 2005, 3.

4. Niigata-ken Sabō Jigyō Hyakunen Kinen Niigata-ken Doboku-bu 1982, map no. 13.

5. Shimura 1995.

6. Andō 1995, 22.

7. These documents are described and analyzed in somewhat greater detail than below in Brown 1999, 161–227.

8. Presenting papers at professional meetings in the United States, I have heard skepticism that redistributions were actually implemented. To many, the idea of spending the weeks necessary to actually reevaluate and measure even a medium-sized village of 200 *koku* seems irrational, as do several other facets of these systems. The hostile response to a presentation on *warichi* by one young Japanese colleague echoes the concerns of my skeptical American colleagues. From this perspective, the following discussion presents evidence of the strong

commitment of Iwade shareholders to redistribution despite its heavy demands on their time.

9. This supposition is supported by the fact that in cases where a family possessed a surfeit of garden land (for cultivating vegetables, generally), that family did not participate in the reallocation of some mountain land. McKean 1985 and 1989 describe similar mechanisms for *iriai*.

10. Funahashi 1995, 204. This resident's land appears to represent *tobichi*, land separated from the village that administered it.

11. Takazawa 1979, 1190.

12. There appear to have been very slight differences in the redistribution shares held and the share of taxes that a family paid. However, there is no indication that this difference somehow altered the basic functioning of *warichi* in Iwade village. See Funahashi 1995, 204.

13. The discussion that follows is based on Funahashi 2004, especially 306–307.

14. See Brown 1987b, 1989, and 1998.

15. Brown 1989, 1999.

16. Within the small area cases, there is frequent repetition from redistribution to redistribution, so that the same notation appears in the same section of the same *wari* in each implementation. For a listing of larger area cases, see Brown 1999, 196–205. Despite the presence of these cases, cases of less than 30 *bu* are considerably more numerous.

17. Documents with titles like "Lot Exchange Notebook" (*kujikae chō*) record exchanges of shares based on drawing lots but without remeasurement. In the full collection of Iwade village notebooks on redistribution, this distinction is hard and fast.

18. See Brown 1999, table 12.

19. An experienced surveyor could expeditiously and accurately sight-measure. See Richeson 1966.

20. A possible exception is discussed under the heading of special exemptions (*hikikuji*) in Chapter 7. There are numerous other examples of a widespread Japanese emphasis on the output of a piece of land rather than on actual measurement. See Brown 1998, 1987a, 1989.

21. In the Hōreki 6–7 notebook Uranaka Number 1 *wari* section 2, the following note appears: "The 100-*bu wari* section 13 notes, 'Shinden 70 *bu* unmeasured.' "

22. Shimonishi *wari* was commonly divided into just four sections, but this instance differs from those I treat here as "fewer than sixteen *wari*."

23. Hōei 7 (1710) 50-*bu wari* has nine, the seventeenth *wari* has four, Residential *wari* has nine, and Sukezawa and Hatake 50-*bu wari* have twelve shares. In the Hōreki 6–7 (1756–1757) redistribution, Leftover *wari* has six and Nakama-batake has six; the Bunka 6 notebook Senshō Sawada 70/Yamasawa Yoshino 30-*bu wari* has six.

24. See document number 8008, the 60-*bu wari*, in the two Bunsei note-books: the Takehara 50-*bu wari* Number 3 and Hatake (Shinden) 60-*bu wari*, the Bunka 6 Shinden-batake 60-*bu wari* and Takehara Hatake 50-*bu wari* Number 1, the Hōreki Residential *wari* (an exceptional case), and the Shinden 60-*bu wari*.

25. Fall 1993 interviews in Kumayama, where a part of the town practiced *warichi*, and several mountain communities in Tottori, near the Okayama-Tottori boundary, that practiced swidden agriculture.

## Chapter 6: *Warichi* and Natural Hazard

1. Anyone who does not believe that climate factors vary even over modest distances should travel along the Niigata coast in winter. Snow precipitation and accumulation patterns differ significantly, even over distances of as short as several kilometers.

2. Cutter 1993, chapter 2.

## Chapter 7: Luck of the Draw? Outcomes and Disputes

1. For one example of this practice, see Niigata-ken Nōgyō Kyōiku Sentaa 1968, 110-114. Niigata-ken Naimubu 1929, 48–49, discusses more examples. For a discussion of the interrelationship between *warichi* and accumulation of land rights by some individuals, see Nōsei Chōsa Kai 1967, 103–116, 163–164. For an example of a listing of salary lands, see "Kyōhō 2 (1717) nen Echigo-kuni Kariwa-kōri Fujii-kumi Tsurugi-mura sashidashi chō," in Ihara Jun'ei monjo.

2. Adding land close to one's house reduced access time to fields. The compensatory loss of superior paddy was a substantial penalty.

3. Funahashi 1995, 236–237.

4. The two 1818 Bunsei notebooks note different numbers of *hikikuji* although they are purportedly for the same redistribution. There are 27 sections within *wari* that are noted as *hikikuji* in the *gechō* (6.7 percent of the 403 sections in the village), while the *oboechō* notes 26 sections out of 420 (6.2 percent). In the *gechō*, almost a third of the shareholders have some right to *hikikuji*. This is a considerably larger share than in the *oboechō*. Why such differences appear cannot presently be determined.

5. In all cases where the holder of *hikikuji* is not specified, the holder appears to be the document author.

6. Appendix B, Tables 3 and 4.

7. From the Hō'ei 1710 and Hōreki 1756–1757 redistributions, residential lands were completely excluded from Iwade's *warichi* records. According to a note in the Bunsei notebooks, some (perhaps a very large part) of these lands were reclassified as being outside the purview of redistribution. The remainder

of residential lands were incorporated in other, existing *wari* or made part of a completely new *wari*. Although not a part of regular exchanges, a consciousness of residential land as a separate category of land remained, and in rare cases one catches a glimpse of that awareness. We see it in remarks about part of a *wari* having to be treated as residential land. For instance, in the Bunsei era 1818 *gechō*, in the record of the second section of the Asabatake 30-*bu* number 1 *wari*, section 2, we read, "Insufficient residential lands, 29 *bu*, at present, 100 *bu* converted to paddy."

8. Funahashi 1995, 216–221.

9. "An'ei 2 mi Ozawa mura Sōzaeimon chi heikin negai narabini utsushi, toriatsukainin kakitsuke," Hoshino-ke monjo VI-10, B5–7, 1521, 4447.

10. Procedures existed for changing mountain land to dry field, or paddy and dry field to paddy, but village approval had to be granted. Conversions were commonly treated as land reclamation.

11. "Ta-hatake chiwari chō, Hōreki roku mi aki ta-wari, Hōreki nana ushi haru hatake-wari san gatsu." Satō-ke monjo 57-A, 8009.

12. See the various documents related to the petitions of Sōzaeimon of Ozawa village in the Hoshino-ke monjo collection: VI-10 B5–7, 1537, 4463; VI-10 B5–7, 1538, 4464; VI-10 B5–7, 1521, 4447.

13. See, for example, "Kansei jū ushi nen yori Ozawa-mura Tomizaeimon deiri sho ikken," Hoshino-ke monjo VI-10, B3–4, 775, 3701; "An'ei hachi nen chi kabu ikken gansho tome chō," Ihara Jun'ei monjo. It is not uncommon for anthropologists to assume that systems like *warichi* require a "closed" village structure, but this was not generally so for Japanese joint ownership.

14. For one example of the way villagers divided up such expenses among themselves, see the Tenpō 7 "Hamen hikimai narabini on rei muki sho nyūyō wappu chō," Kawaji kuyū monjo C-1-148.

15. See the pledge of Sōzaeimon of Ozawa village, An'ei 6.2.12, Hoshino-ke monjo VI-10, B5–7, 1538, 4464, for one example.

16. See note 14 above for sources.

17. Fukaya 1969, 223–243.

## Chapter 8: Adaptability, Survivability, and Persistent Influences

1. "Aisadame mōsu chiwari renban shōmon no koto," Genbun 2 (1790), document 66, Tokamachi-shi Shi Hensanshitsu 1992–1997, *Shiryō hen* 4, *Kinsei* 1, 745.

2. On the hurried pace of land surveys, see Brown 1993, chapter 3; for routine errors in the land survey process, see Brown, 1987b, 115–155.

3. Document 68, "Osorenagara kakituke wo motte negaiage tatematsuri sōrō no koto," Tokamachi-shi Shi Hensanshitsu 1992–1997, *Shiryō hen* 4, *Kinsei* 1, 746–748.

4. Matsuzawa monjo, D-3, 225, "Hitofuda no koto," Tenmei 6 (1786).

5. Sakai-ke monjo, no. 3452, "Echigo-kuni Uonuma-kōri Minami Abusaka–mura shinden kenchi chō," Bunka 10 (1813), and no. 3454, "Echigo-kuni Uonuma-kōri Kita Abusaka–mura shinden kenchi chō," Bunka 10 (1813).

6. Saiki Shigemasa monjo, no. 206-356, "Wadan no tame torikae shōmon no koto," Kyōwa 3 (1803).

7. Shimada Keiichi monjo, no. 2-507, "Torikae mōsu tame yakutei shōmon no koto," Ansei 2 (1855).

8. This purpose was also present in a number of reclamation projects. Some domains aggressively rounded up landless and poor farmers as cultivators on large reclamation projects. See, for example, Wakabayashi 1962.

9. Tokamachi Shimada-ke monjo, no. 2-504, "Torikae mōsu tame yakutei shōmon no koto," Ansei 2 (1855).

10. Shimada Keiichi monjo, no. 2-765, "Mura chū iriai-chi taidan chō," Meiji 8 (1875).

11. Fukushima 1968, 51–54, for example.

12. Inoue 1970, 217, indicates that the Meiji land tax reforms evoked little opposition. Data on peasant protests gathered by White 1995 for his massive computer data base on *ikki* show no rural events in Niigata during the period of the Meiji land tax reforms. Aoki 1968 shows three rural events in Meiji 5, the year in which the plans for reform were first announced (appendix, p. 36). All occurred before the publication of tax reforms. Another incident (p. 43) took place the following year in a dispute between villages over access to common lands *(iriai-chi)*. Aoki (1971, appendix, p. 34, and 1968, 36) indicates that Niigata's pattern of disturbances prior to land tax reform followed by a sharp decline in numbers thereafter was not unusual.

13. Aono 1977, 105.

14. Niigata-ken Nōgyō Kyōiku Sentaa 1968, 48–52. For other examples, see pp. 102–103, 216, 22, 248–249, and 252.

15. Ibid., 236–239.

16. See Niigata-ken Nōchika 1957, 4:167, 179, for one case.

17. Niigata-ken Nōgyō Kyōiku Sentaa 1968, 178, 200, and 236.

18. Ibid., 39–41.

19. Ibid., 59.

20. Smith 1959 outlines this pattern.

21. In this sense their position as "absentee" landlords was very different from that of later large Niigata landlords, who often lived in Tokyo after the turn of the century, severing day-to-day ties to communities where they held lands.

22. Matsuzawa monjo, no. 41, "Chikei yuzuri shōmon no koto," Tenmei 6 (1786).

23. "Osorenagara kakitsuke wo motte negaiage tatematsuri sōrō no koto," Kansei 2 (1790) no. 70, Tokamachi-shi Shi Hensanshitsu 1992–1997, *Shiryō hen* 4, *Kinsei* 1, 753–754.

24. Matsuzawa monjo, no. 571, "Ai mōshi watashi hito fuda no koto," Tenmei 6 (1786).

25. Quoted in Aono 1977, 108.

26. Niigata-ken Nōgyō Kyōiku Sentaa 1968, 39–41, 216; Niigata-ken Naimubu 1929, 248–249, 254–260, 271–280.

27. Niigata-ken Naimubu 1929, 281.

28. Ibid., 260–272. Data are presented in two tables; rent changes are noted in the "remarks."

29. Ibid., 250–251.

30. Ibid., 248.

31. Limited evidence exists for the presence of joint ownership in Kita Kanbara, and none has been found to date for Iwafune-gun.

32. Nōsei Chōsa Kai 1967, 164. Although not a majority practice, it was not uncommon for nonresidents to serve as headmen in villages other than that in which they resided. Furthermore, one could be appointed to head a group of villages.

33. Ibid., 115–116. This same pattern of accumulation is evident in another large landholder, the Haradamaki family. The much smaller Satō family followed a similar pattern, but it began to sell off its holdings before the end of the Tokugawa era. Kokuritsu Bungaku Kenkyū Shiryō-kan 1983–1991, 1:284.

34. See, for example, Pratt 1999. The loss of office was in part a result of the Meiji consolidation of local administration.

35. Nōsei Chōsa Kai 1967, 188.

36. Niigata-ken Naimubu 1929, 253.

37. Ibid., 233.

38. Ibid., 280.

39. Niigata-ken Nōgyō Kyōiku Sentaa 1968, 53. The village's commonly held lands continued to be rotated for almost two decades thereafter.

## Chapter 9: Final Reflections

1. Other complex forms of tenure can be found. See Fukushima 1968.

2. For example, Barzel 1989, 1997.

3. For example., Fukuda 2005.

4. Barzel 1989.

5. For a very brief introduction to Okinawan *warichi*, the case of Kudaka-jima, see Brown 2002b, 47.

6. Postwar Nagaoka *warichi*, for example, also met this fate.

7. On these issues, see McKean 1985 and McKean 1989. See also Totman 1984; Harada 1969.

8. Harada 1969 is a classic study of the destruction of commons under privatizing pressures.

9. From the perspective of the less than enthusiastic participants, the at-

titude might have been that redistributive practices were not such a burden as to be worth fighting to abandon.

10. The head of a Nagaoka family spoke to me in 1994 about pressure tactics he used to get cooperation on some issues when the chair of the local agricultural committee proved obstinate.

11. Simmel (1964) observed that what makes a community is not the absence of conflict but agreement on the rules for dealing with conflict, determining which conflicts and methods of pursuing conflict were tolerable and which were not.

12. Tilly 1985, 169–191.

13. The domestic confrontations were the battles of Sekigahara (1600), the two Osaka campaigns of 1614–1615 that secured Tokugawa dominance by eliminating Hideyoshi's heir, and the Shimabara Revolt (1637). The two international conflicts were Hideyoshi's invasions of Korea in the 1590s.

14. Vaporis 2008.

15. Yamamura 1974 outlines trends in samurai income and the secular value of their stipends.

16. Brown 1987a and 1988.

17. Miller (1992) makes this point as part of his more generalized argument.

18. Jack Hirschleifer as quoted in Thaler 1992, 9.

19. White 1995, 151, 153.

20. Thaler 1992, chapter 2, "Cooperation," summarizes the experiments that led to these insights.

21. Brown 1997.

22. Ono 1931, 30. Lieban 1956 is largely focused on *warichi* on Kudakajima.

23. Access to political power was by no means automatic, and a number of disputes in eighteenth- and nineteenth-century villages focused on the refusal of established village elites to share power with emerging, *nouveaux riches* families. See Smith 1959.

24. Ono 1924 presents one classic study.

25. On *kaimodoshi* in Echigo, see Matsunaga 1989, 89–92. Matsunaga has often discussed with me the general trends in the hundreds of land "title" documents he has encountered over the five decades of his research on central and southern Echigo. On divided ownership rights in a modern context, see Barzel 1989.

26. Hayami 2004, 9, summarizes Heilbronner's views.

27. Simmel (1964) defines "communities" in this way.

28. Kobayashi Torasaburo chose to invest relief rice in local education to promote prosperity rather than to distribute it all immediately. Kobayashi's decision is celebrated in Yamamoto Yuzo's play *Kome hyappyō* (for an English translation, see Yamamoto 2001).

# Bibliography

## Manuscript Collections

Echigo-kuni Kubiki Kōri Iwade-mura Satō-ke monjo. Kokuritsu Bungaku Kenkyū Shiryōkan, Tokyo.

Hoshino-ke monjo. Yoshikawa Chōshi shiryō. Yoshikawa Chōshi Hensanshitsu, Yoshikawa-machi, Niigata Prefecture.

Ihara Jun'ei monjo. Oaza Tsurugi, Kashiwazaki-shi, Niigata Prefecture.

Kawaji kuyū monjo. Tokamachi-shi Shi Hensan Shitsu, Tokamachi, Niigata Prefecture.

Matsuzawa-ke monjo. Tokamachi-shi Shi Hensanshitsu, Tokamachi, Niigata Prefecture.

Saiki Shigemasa monjo. Tokamachi-shi Shi Hensanshitsu, Tokamachi, Niigata Prefecture.

Sakai-ke monjo, Tokamachi-shi Shi Hensanshitsu, Tokamachi, Niigata Prefecture.

Shimada Keiichi monjo. Tokamachi-shi Shi Hensanshitsu, Tokamachi, Niigata Prefecture.

Tokamachi Shimada-ke monjo. Tokamachi-shi Shi Hensanshitsu, Tokamachi, Niigata Prefecture.

## Printed Sources

Unless otherwise noted, the place of publication for Japanese-language works is Tokyo.

Ackerman, Bruce A., ed. 1975. *Economic Foundations of Property Law*. Boston: Little, Brown Company.

Andō Masahito. 1995. "Echigo-kuni Kubiki-gun Iwade-mura Satō-ke monjo no kōzō." In Watanabe Naoshi, ed., *Kinsei beisaku tansaku chitai no sonraku shakai: Echigo-kuni Iwade-mura Satō-ke monjo no kenkyū*, 1–53. Iwata Shoin.

Aoki Koji. 1968. *Ikki, Meiji nōmin sōgō no nenjiteki kenkyū*. Shinseisha.

———. 1971, *Hyakushō ikki sōgō nenpyō*. San'ichi Shobo.

Aono Shunsui. 1977. "Echigo ni okeru warichi sei." *Hiroshima Daigaku Kyōiku Gakubu kiyō, dai 2 bu*, no. 26, 101–112.

———. 1982. *Nihon kinsei warichi sei shi no kenkyū*. Ozankaku Shuppan.

Araki Moriaki. 1959. *Bakuhan taisei shakai no seiritsu to kōzō*. Yūhikaku.

Arashi Keiichi. 1975. *Kinsei inasaku gijutsu shi*. Nōbunkyō.

Arimoto Masao. 1967. *Chisō kaisei to nomin tōsō*. Shinseisha.

Asao Naohiro et al. 1996. *Kadokawa shinpan Nihon shi jiten*. Kadokawa Shoten.

Barzel, Yoram. 1989 (2nd ed., 1997). *Economic Analysis of Property Rights*. Cambridge, UK: Cambridge University Press.

Befu, Harumi. 1968. "Village Autonomy and Articulation with the State." In John W. Hall and Marius B. Jansen, eds., *Studies in the Institutional History of Early Modern Japan*, 301–314. Princeton: Princeton University Press.

Berkes, Fikret. 1980. "Marine Inshore Fishery Management in Turkey." In *Proceedings of the Conference on Common Property Resource Management*, 63–83. Washington, DC: National Research Council.

Berry, Mary Elizabeth. 1986. "Public Peace and Private Attachment: The Goals and Conduct of Power in Early Modern Japan." *Journal of Japanese Studies* 12:2 (Summer), 237–271.

Bird, Isabella. 1984 (1880). *Unbeaten Tracks in Japan*. Boston: Beacon Press.

Blackford, Mansel G. 1988. *The Rise of Modern Business in Great Britain, the United States, and Japan*. Chapel Hill: University of North Carolina Press.

Bray, Francesca. 1986. *The Rice Economies: Technology and Development in Asian Societies*. Oxford: Basil Blackwell.

Broadbent, Seymour A. 1966. *Industrial Dualism in Japan: A Problem of Economic Growth and Structural Change*. Chicago: Aldine Press.

Brown, Philip C. 1987a. "Land Redistribution Schemes in Tokugawa Japan: An Introduction." *Occasional Papers of the Virginia Consortium for Asian Studies*, vol. 4 (Spring), 35–48.

———. 1987b. "The Mismeasure of Land: Land Surveying in the Tokugawa Period." *Monumenta Nipponica* 42:2 (Summer), 115–155.

———. 1988. "Practical Constraints on Early Tokugawa Land Taxation: Annual Versus Fixed Assessments in Kaga Domain." *Journal of Japanese Studies* 14:2 (Summer), 369–401.

———. 1989. "Never the Twain Shall Meet: European and Japanese Land Survey Techniques in Tokugawa Japan." *Chinese Science* 9, 53–79.

———. 1993. *Central Authority and Local Autonomy in the Formation of Early Modern Japan: The Case of Kaga Domain*. Stanford: Stanford University Press.

———. 1997. "State, Cultivator, Land: Determination of Land Tenures in Early Modern Japan Reconsidered." *Journal of Asian Studies* 56:2 (May), 421–444.

———. 1998. "A Case of Failed Technology Transfer—Land Survey Technology in Early Modern Japan." *Senri Ethnological Studies* 46 (March), 83–97.

———. 1999. "Warichi seido: soto kara mita omoshirosa, naka kara mita fukuzatusa." *Shiryōkan kenkyū kiyō* (Kokuritsu Bungaku Kenkyūjo Shiryōkan), March, 161–227.

———. 2002a. "Essays on the State of the Field: An Introduction." *Early Modern Japan: An Interdisciplinary Journal* 10:1 (Spring), 1–3.

———. 2002b. "Harvests of Chance: Corporate Control of Arable Land in Early Modern Japan." In John F. Richards, ed., *Land, Property, and the Environment*, 38–70. Oakland, CA: ICS Press.

———. 2003a. "The Political and Institutional History of Early Modern Japan." *Early Modern Japan: An Interdisciplinary Journal* 11:1 (Spring), 3–30.

———. 2003b. "Summary of Discussions: The State of the Field in Early Modern Japanese Studies." *Early Modern Japan: An Interdisciplinary Journal* 11:1 (Spring), 54–63.

———. 2005. "Corporate Land Tenure in Nineteenth-Century Japan: A GIS Assessment." *Historical Geography* 33 (May), 99–117.

———. 2006. "Tochi warikae sei to shizen kankyo." In Hara Naofumi and Ohashi Kōji, eds., *Nihonkai iki rekishi taikei*, 177–207. Osaka: Seibundō Shuppan.

———. 2008a. "Constructing Nature in Japan." Paper presented at the Japan's Natural Legacies conference, Big Sky, Montana, October 4, 2008.

———, ed. 2008b. "Exploring Historical Space and Environments in the History/Social Studies Classroom." *Journal of the Association for History and Computing* 11:2 (August). http://quod.lib.umich.edu/cgi/t/text/text-idx ?c=jahc;view=text;rgn=main;idno=3310410.0013.102.

Brown, Philip C., and Taniguchi Shinko. 2000. "Amerika ni okeru Nihon kinsei-shi kenkyū no dōkō." *Nihonshi kenkyū* 453 (May), 53–70.

Brubaker, Earl R. 1975. "Free Ride, Free Revelation, or Golden Rule?" *Journal of Law and Economics* 18:1 (April), 147–161.

Cutter, Susan L. 1993. *Living with Risk*. London: Edward Arnold Press.

Dahlman, Carl. 1980. *The Open Field System and Beyond: Property Rights Analysis of an Economic Institution*. Cambridge, UK: Cambridge University Press.

Davis, David L. 1974. "Ikki in Late Medieval Japan." In John W. Hall and Jeffrey P. Mass, eds., *Medieval Japan*, 221–247. New Haven: Yale University Press.

Edwards, Walter. 1991. "Buried Discourse: The Toro Archaeological Site and

Japanese National Identity in the Early Postwar Period." *Journal of Japanese Studies* 17:1 (Winter), 1–23.

Francks, Penelope. 2006. *Rural Economic Development in Japan: From the Nineteenth Century to the Pacific War.* London: Routledge.

Fukaya Katsumi. 1969. "Kansei-ki no Tōdō han." *Mie-ken kyōdo shiryō sōsho 17, Tsū-shi*, 223–243. Mie-ken: Mie-ken Kyōdo Shiryō Kankōkai.

Fukuda Chizuru. 2005. *Oie sōdō: daimyō-ke wo yurugashita kenryoku tōsō.* Chūō Kōron Shinsha.

Fukushima Masao. 1968. *Chisō kaisei.* Yoshikawa Kōbunkan.

Funahashi Akihiro. 1995. "Sonraku kōzō to sono hen'yō: warichi to kosakuchi kei'ei wo megutte." In Watanabe Naoshi, ed., *Kinsei beisaku tansaku chitai no sonraku shakai: Echigo-kuni Iwade-mura Satō-ke monjo no kenkyū*, 201–260. Iwata Shoin.

———. 2004. *Kinsei jinushisei to chiiki shakai.* Iwata Shoin.

Furushima Toshio. 1939. "Warichi seido ni kansuru bunken." *Nōgyō keizai kenkyū* 16:4, 134–162.

———. 1961. "Nōgyō no hatten—inasaku o chūshin ni." In Chihōshi Kenkyū Kyōgikai, ed., *Nihon sangyō shi taikei*, vol. 1, *Sōron hen*, 87–131. Tokyo Daigaku Shuppan Kai.

———. 1975. *Nihon nōgyō gijutsu shi.* Tokyo Daigaku Shuppan Kai.

Hakui-shi Shi Hensan Iinkai. 1975. *Hakui-shi shi. Kinsei hen.* Kanazawa: Hakui-shi Yakusho.

Hall, John W. 1981. "Hideyoshi's Domestic Policies." In John W. Hall, Nagahara Keiji, and Kozo Yamamura, eds., *Japan before Tokugawa: Political Consolidation and Economic Growth, 1500–1650*, 194–223. Princeton: Princeton University Press.

Hanawalt, Barbara. 2007. *The Wealth of Wives: Women, Law, and Economy in Late Medieval London.* Oxford: Oxford University Press.

Hanley, Susan B., and Kozo Yamamura. 1977. *Economic and Demographic Change in Preindustrial Japan, 1600–1868.* Princeton: Princeton University Press.

Hanpō Kenkyū Kai, ed. 1966. *Hanpō shū*, vol. 4: *Kanazawa han.* Sobunsha.

Hara Shogo. 1971. "Kaga-han no denchiwari seido ni tsuite." *Tokugawa rinsei shi kenkyūjo kenkyū kiyō*, March, 279–302.

Harada Toshimaru. 1969. *Kinsei iriai seido kaitai katei no kenkyū: Yamawari seido no hassei to sono henshitu.* Hanawa Shobō.

Hardin, Garrett. 1968 (1975 reprint). "The Tragedy of the Commons." *Science* 162. Reprinted in Bruce A. Ackerman, ed., *Economic Foundations of Property Law*, 2–11. Boston: Little, Brown Company.

———. 1991. "The Tragedy of the Unmanaged Commons: Population and the Disguises of Providence." In Robert V. Andelson, ed., *Commons without Tragedy: Protecting the Environment from Overpopulation—a New Approach*, 162–185. London: Shepheard-Walwyn.

Hauser, William B. 1985. "Osaka Castle and Tokugawa Authority in Western

Japan." In Jeffrey P. Mass and William B. Hauser, eds., *The Bakufu in Japanese History*, 153–172. Stanford: Stanford University Press.

Hayami, Akira. 2004. "Introduction." In Akira Hayami, Osamu Saito, and Ronald Toby, eds., *The Economic History of Japan: 1600–1990*, vol. 1: *Emergence of Economic Society in Japan: 1600–1859*, 1–35. Oxford: Oxford University Press.

Hayami Akira et al. 1988–1990. *Nihon keizai-shi.* 8 vols. Iwanami Shoten.

Hayami, Akira, Osamu Saito, and Ronald Toby, eds. 2004. *The Economic History of Japan: 1600–1990*, vol. 1: *Emergence of Economic Society in Japan: 1600–1859.* Oxford: Oxford University Press.

Heki Ken, ed. 1970. *Kaga-han shiryō.* 18 vols. Osaka: Seibundo Shuppan.

———. 1973. *Kano kyōdo jii.* Meicho Shuppan.

Henderson, Dan Fenno, comp. 1975. *Village "Contracts" in Tokugawa Japan: Fifty Specimens with English Translations and Comments.* Seattle: University of Washington Press.

Howell, David. 1995. *Capitalism from Within: Economy, Society, and the State in a Japanese Fishery.* Berkeley: University of California Press.

Inoue Toshio. 1970. *Niigata-ken no rekishi.* Kenshi shireezu 15. Yamakawa Shuppansha.

Ishii Shirō. 1966. *Nihon kokusei shi kenkyū: kenryoku to tochi shoyū.* Tokyo Daigaku Shuppan Kai.

Jansen, Marius B. 1968. "Tosa in the Seventeenth Century: The Establishment of Yamauchi Rule." In John W. Hall and Marius B. Jansen, eds., *Studies in the Institutional History of Early Modern Japan*, 112–130. Princeton: Princeton University Press.

———. 2000. *The Making of Modern Japan.* Cambridge, MA: Belknap Press.

Jōetsu-shi Shi Hensan Iinkai. 1998. *Kuwadoridani bunka no kiroku.* Jōetsu-shi Shi Sosho 3. Jōetsu-shi, Niigata: Jōetsu-shi.

———. 1999–2004. *Jōetsu-shi shi.* 21 vols. Jōetsu-shi, Niigata: Jōetsu-shi.

Judd, Richard. 1997. *Common Lands, Common People: The Origins of Conservation in Northern New England.* Cambridge, MA: Harvard University Press.

Kamikaji-ke Monjo Chōsa Dan, ed. 1977. *Kamikaji-ke monjo.* Kanazawa: Ishikawa-ken Kenritsu Toshokan.

Kanai Madoka, ed. 1985. *Dokai kōshū ki.* Shin Jinbutsu Ōraisha.

Kawata Shirō. 1912. "Nōmin tochi aichakushin reikyaku no keikō." *Keizai ronsō* 17:6 (December), 70–86.

Kikuchi Toshio. 1977. *Shinden kaihatsu kaitei zōhōban.* Kokon Shoin.

Keirstead, Thomas. 1992. *The Geography of Power in Medieval Japan.* Princeton: Princeton University Press.

Kelly, William W. 1982. *Water Control in Tokugawa Japan: Irrigation Organization in a Japanese River Basin, 1600–1870.* East Asian Papers no. 31. Ithaca: Cornell China-Japan Program.

Knowles, Anne Kelly, ed. 2002. *Past Time, Past Place: GIS for History.* Redlands, CA: ESRI Press.

———, ed. 2005. "Thematic Issue: Emerging Trends in Historical GIS." *Historical Geography* 33.

———, ed. 2008. *Placing History: How Maps, Spatial Data, and GIS Are Changing Historical Scholarship.* Digital supplement edited by Amy Hillier. Redlands, CA: ESRI Press.

Kokudo Chiriin. 1982. *National Atlas of Japan.* CD. Kokudo Chiriin.

Kokuritsu Bungaku Kenkyū Shiryōkan 1983–1991. *Echigo-kuni Kubiki-gun Iwate-mura Satō-ke monjo mokuroku.* 4 vols. Kokuritsu Bungaku Kenkyū Shiryōkan.

Kondō Tetsuo. 1967. *Chisō kaisei no kenkyū: jinushisei to no kanren ni oite.* Miraisha.

Kumayama Chōshi Hensan Iinkai. 1993. *Kumayama chōshi, Ōaza hen.* Kumayama-machi. Okayama: Kumayama-machi.

Kurosaki Chōshi Hensan Iinkai. 1994–2000. *Kurosaki chōshi.* 8 vols. Niigata, Kurosaki: Kurosaki-machi.

Lewis, Kären Wigen. 1985. "Common Losses: Transformations of Commonland and Peasant Livelihood in Tokugawa Japan, 1603–1868." M.A. thesis, University of California, Berkeley.

Libecap, Gary. 1989. *Contracting for Property Rights.* Cambridge, UK, and New York: Cambridge University Press.

Lieban, Richard. 1956. "Land and Labor on Kudaka Island." Ph.D. thesis, Columbia University.

Maki Chōshi Hensan Iinkai. 1988–1994. *Maki chōshi.* 8 vols. Maki-machi, Niigata: Maki-machi.

Makino Shinnosuke. 1928. *Buke jidai shakai no kenkyū.* Toe Shoin.

Matsunaga Yasuo. 1989. *Kinsei nōson shi no kenkyū: kinsei zenki Echigo ni okeru nōson kōzō to nōgyō keiei.* Kyoto: Hōritsu Bunka Sha.

Matsushita Shirō. 1984. *Bakuhansei shakai to kokudakasei.* Hanawa Shobo.

McClain, James. 2001. *Japan: A Modern History.* New York: W. W. Norton.

McCloskey, Donald [a.k.a. Deirdre] N. 1975a. "The Economics of Enclosure: A Market Analysis." In William N. Parker and Eric L. Jones, eds., *European Peasants and Their Markets*, 123–160. Princeton: Princeton University Press.

———. 1975b. "The Persistence of English Common Fields." In William N. Parker and Eric L. Jones, eds., *European Peasants and Their Markets*, 73–119. Princeton: Princeton University Press.

McKean, Margaret A. 1985. "The Japanese Experience with Scarcity: Management of Traditional Common Lands." In Kendall E. Bailes, ed., *Environmental History: Critical Issues in Comparative Perspective*, 334–359. Lanham, MD.: University Press of America.

———. 1989. "Management of Traditional Common Lands (*iraichi*) in Japan."

In Panel on Common Property Resource Management, Board on Science and Technology for International Development, Office of International Affairs, National Research Council, *Proceedings of the Conference on Common Property Resource Management*, 533–589. Washington, DC: National Academy Press. A revised and updated version appears in Daniel Bromley, David Feeny, Margaret A. McKean, Pauline Peters, Jere Gilles, Ronald Oakerson, C. Ford Runge, and James Thomson, eds., *Making the Commons Work: Theory, Practice, and Policy* (1992), 63–98. San Francisco: Institute of Contemporary Studies.

Miller, Gary. 1992. *Managerial Dilemmas: The Political Economy of Hierarchy*. Cambridge, UK: Cambridge University Press, 1992.

Mitsuke-shi Shi Hensan Iinkai. 1976–1984. *Mitsuke-shi shi*. 7 vols. Mitsuke-shi, Niigata: Mitsuke-shi.

Miyagawa Mitsuru. 1959. *Taikō kenchi ron*. Ochanomizu Shobō.

Miyagawa, Mitsuru, with Cornelius Kiley. 1977. "From Shōen to Chigyō." In John W. Hall and Toyoda Takeshi, eds., *Japan in the Muromachi Age*, 89–105. Berkeley: University of California Press.

Miyamoto, Matao. 2004. "Quantitative Aspect of Tokugawa Economy." In Hayami et al., eds., *The Economic History of Japan: 1600–1990, vol. 1: The Emergence of Economic Society: 1600–1859*, 36–84. Oxford: Oxford University Press.

Nagahara, Keiji, with Kozo Yamamura. 1977. "Village Communities and Daimyo Power." In John W. Hall and Toyoda Takeshi, eds., *Japan in the Muromachi Age*, 116–121. Berkeley: University of California Press.

Nagaoka-shi Shi Hensan Iinikai. 1993–1998. *Nagaoka-shi shi*. 8 vols. Nagaoka-shi, Niigata: Nagaoka-shi.

Nakada Kaoru. 1904. "Echigo-kuni warichi seido." *Kokka gakkai zasshi* 18:205 (March), 51–76, and 18:206 (April), 27–63.

Nakazato Sonshi Senmon Iinkai, ed. 1985–1988. *Nakazata sonshi*. 4 vols. Nakazato-mura, Naka Kubiki–gun, Niigata-ken: Nakazato-mura Kyōiku Iinkai.

Nihon Kasen Kyōkai. 2005. *Kasen binran 2004*. Kokudo Kaihatsu Chōsakai.

Niigata-ken. 2005. "Kakizaki kawa suikei seibi kihon hōshin." March, p. 3. PDF file available at http://www.pref.niigata.jp/doboku/engawa/sosiki/honcho/kak/kak_pdf/Kakizaki_hou.pdf (accessed November 1, 2005).

Niigata-ken Naimubu. 1929. *Niigata-ken ni okeru warichi seido*. Niigata: Niigata-ken Naimubu.

Niigata-ken Nōchika, ed. 1957. *Niigata-ken nōchi kaikaku shi shiryō*. 6 vols. Niigata: Niigata-ken Nōchi Kaikakushi Kankōkai.

Niigata-ken Nōgyō Kyōiku Sentaa. 1968. *Niigata-ken Nishi Kanbara-gun ni okeru warichi seido no chōsa—nōminteki tochi shoyū no rekishiteki tenkai*. Niigata: Niigata Kenritsu Kōnōkan Kōtōgakkō.

Niigata-ken Sabō Jigyō Hyakunen Kinen Niigata-ken Doboku-bu. 1982. *Jisu-*

*beri chikei kūchū shashin ni yoru handoku zu.* Map no. 13. Niigata: Niigata-ken Sabō Jigyō Hyakunen Kinen Niigata-ken Doboku-bu.

Niigata-ken Shi Hensanshitsu. 1990. *Niigata no ayumi: Niigata kenshi gaisetsu.* Niigata.

Niigata-shi Shi Hensan Iinkai. 1989–1998. *Niigata-shi shi.* 19 vols. Niigata: Niigata-shi.

North, Douglas C. 1990. *Institutions, Institutional Change, and Economic Performance.* Cambridge, UK: Cambridge University Press.

North, Douglas C., and Robert Paul Thomas. 1973. *The Rise of the Western World: A New Economic History.* Cambridge, UK: Cambridge University Press.

Nōsei Chōsa Kai. 1967. *Imai-ke no jinushi kōzo. Niigata-ken dai jinushi shozo shiryō 9.* Niigata: Nōsei Chōsa Kai.

Oda Kichinojo. 1929. *Kaga-han nōsei shi kō.* Toe Shoin.

Oishi Tsunetaka. 1969. *Jikata hanrei roku. Jō, ge.* Kondō Shuppansha.

Ono Takeo. 1924. *Eikosakuron.* Ganshodo Shoten.

———. 1931. *Tochi keizai shi kōshō.* Ganshodo Shoten.

Ooms, Herman. 1996. *Tokugawa Village Practice.* Berkeley: University of California Press.

Ostrom, Elinor. 1990. *Governing the Commons: The Evolution of Institutions for Collective Action.* Political Economy of Institutions and Decisions. Cambridge, UK: Cambridge University Press.

Pallot, Judith. 1999. *Land Reform in Russia, 1906–1917: Peasant Responses to Stolypin's Project of Rural Transformation.* Oxford: Oxford University Press.

Panel on Common Property Resource Management, Board on Science and Technology for International Development, Office of International Affairs, National Research Council. 1986. *Proceedings of the Conference on Common Property Resource Management.* Washington, DC: National Academy Press. Many of the essays in this volume were published in revised form in Daniel Bromley, David Feeny, et al., eds., *Making the Commons Work* (1992). San Francisco, CA: Institute of Contemporary Studies.

Popkin, Samuel. 1979. *The Rational Peasant: The Political Economy of Rural Society in Vietnam.* Berkeley: University of California Press.

Pratt, Edward. 1999. *Japan's Protoindustrial Elite: The Economic Foundations of the Gōnō.* Cambridge, MA: Harvard University Asia Center.

Purnell, Jennie. 1999. "With All Due Respect: Popular Resistance to the Privatization of Communal Lands in Nineteenth Century Michoacán." *Latin American Research Review* 34:1, 85–121.

Ravina, Mark. 1999. *Land and Lordship in Early Modern Japan.* Stanford: Stanford University Press.

Richards, John H., ed. 2002. *Land, Property and the Environment.* Oakland, CA: ICS Press.

Richeson, A. W. 1966. *English Land Measuring to 1800: Instruments and Practices.* Cambridge, MA: MIT Press.

Roberts, Luke S. 1998. *Mercantilism in a Japanese Domain: The Merchant Origins of Economic Nationalism in Eighteenth-Century Tosa.* Cambridge, UK: Cambridge University Press.

Saitō, Osamu. 2000. "Marriage, Family Labour and the Stem Family Household: Traditional Japan in a Comparative Perspective." *Continuity and Change* 15:1, 17–45.

Sakai, Robert. 1968. "The Consolidation of Power in Satsuma-han." In John W. Hall and Marius B. Jansen, eds., *Studies in the Institutional History of Early Modern Japan*, 131–139. Princeton: Princeton University Press.

Sasaki Junnosuke. 1964. *Bakuhansei kokka ron.* Tōkyō Daigaku Shuppankai.

Schwarzwalder, Brian. 2000. "China's Farmers Need Long-Term Land Tenure Security, Not Land Readjustment." *Beyond Transition: The Newsletter about Reforming Economies* (World Bank Group), August/September, 21–23. http://www.worldbank.org/html/prddr/trans/augsepoct00/pages21-23.htm.

Schwarzwalder, Brian, Roy Prosterman, Ye Jianping, Jeffrey Riedinger, and Li Pin. 2002. "An Update on China's Rural Land Tenure Reforms: Analysis and Recommendations Based upon a Seventeen-Province Survey." *Journal of Asian Law* 16:1 (Fall), 141.

Scott, James C. 1976. *The Moral Economy of the Peasant: Rebellion and Subsistence in Southeast Asia.* New Haven: Yale University Press.

Shimura Hiroshi. 1995. "Echigo jinushi chitai no ōjoyasei shihai." In Watanabe Naoshi, ed., *Kinsei beisaku tansaku chitai no sonraku shakai: Echigokuni Iwade-mura Satō-ke monjo no kenkyū*, 55–103. Iwata Shoin.

Siebert, Loren. 2000. "Urbanization Transition Types and Zones in Tokyo and Kanagawa Prefectures." *Geographical Review of Japan*, Series B, 73:2, 207–224.

Simmel, Georg. 1964. *Conflict and the Web of Group Affiliations.* New York: Free Press.

Smith, Thomas C. 1959. *The Agrarian Origins of Modern Japan.* Stanford: Stanford University Press.

———. 1968. "The Japanese Village in the Seventeenth Century." In John W. Hall and Marius B. Jansen, eds., *Studies in the Institutional History of Early Modern Japan*, 263–282. Princeton: Princeton University Press.

———. 1970. "Okura Nagatsune and the Technologists." In Albert M. Craig and Donald H. Shively, eds., *Personality in Japanese History*, 127–154. Berkeley: University of California Press.

———. 1973. "Pre-Modern Economic Growth: Japan and the West." *Past and Present* 60, 127–160.

Solnick, Steven L. 1998. *Stealing the State: Control and Collapse in Soviet Institutions.* Cambridge, MA: Harvard University Press.

Stiglitz, Joseph. 1974. "Incentives and Risk Sharing in Sharecropping." *Review of Economic Studies* 41 (April), 219–255.

Suetsugi Tadashi. 2005. *Zukai zatsugaku: kasen no kagaku.* Natsumesha.

Sugimoto Tsutomu. 1976. "Jiseiteki kagaku bunka = jitsugaku no kōryū." In *Taikei Nihon shi sōsho 19, Kagaku shi,* 154–161. Yamakawa Shuppansha.

Takagi, Masao. 2000. "Landholdings and the Family Life Cycle in Traditional Japan." *Continuity and Change* 15:1, 47–75.

Takazawa Yūichi. 1967. *"Warichi* seido to kinseiteki sonraku." *Kanazawa Daigaku keizai ronshū* 6, 125–142.

———. 1979. "Jinushi keiseiki no kosaku kei'ei ni tsuite." In Dokushikai Sōritsu Gojū Nen Kinen, *Kokushi ronshū,* 1189–1208. Kyoto: Dokushikai.

Takayanagi Mitsutoshi and Takeuchi Rizō. 1974. *Kadokawa Nihon-shi jiten,* Kadokawa Shoten.

Teradomari Chōshi Hensan Iinkai. 1988–1992. *Teradomari chōshi.* 6 vols. Teradomari-machi, Niigata: Teradomari-machi.

Thaler, Richard H. 1992. *The Winner's Curse: Paradoxes and Anomalies of Economic Life.* Princeton: Princeton University Press.

Tilly, Charles. 1985. "War Making and State Making as Organized Crime." In Peter B. Evans, Dietrich Rueschemeyer, and Theda Skocpol, eds., *Bringing the State Back In,* 169–191. Cambridge, UK: Cambridge University Press.

Tochinai Reiji. 1936. *Kyū Kaga-han denchiwari seido.* Mibu Shoin.

Tochio-shi Shi Hensan Iinkai. 1974–1981. *Tochio-shi shi.* 32 vols. Tochio-shi, Niigata: Tochio-shi.

Tokamachi-shi Shi Hensanshitsu. 1992–1997. *Tokamachi-shi shi.* 14 vols. Tokamachi-shi, Niigata: Tokamach-shi.

Tonami-shi Shi Hensan Iinkai. 1965. *Tonami-shi shi.* Tonami-shi, Toyama: Tonami-shi Yakusho.

Totman, Conrad. 1984. *The Origins of Japan's Modern Forests: The Case of Akita.* Asian Studies at Hawai'i no. 31. Honolulu: University of Hawai'i Press.

———. 1989. *Green Archipelago: Forestry in Pre-Industrial Japan.* Berkeley: University of California Press.

———. 1993. *Early Modern Japan.* Berkeley: University of California Press.

Toyama-ken Shi Hensan Iinkai, ed. 1974–1985. *Tonami-shi shi.* Toyama: Toyama-ken.

Trewartha, Glenn T. 1965. *Japan, a Geography.* Madison: University of Wisconsin Press.

Tsang, Carol. 2007. *War and Faith: Ikkō Ikki in Late Muromachi Japan.* Cambridge, MA: Harvard East Asian Monographs, Harvard University Asia Center.

Tsubame-shi Shi Hensan Iinkai. 1988. *Tsubame-shi shi.* 4 vols. Tsubame-shi, Niigata; Tsubame-shi.

Tsukamoto Tomohiro. 1992a. "Bakumatsu ki ni okeru warichi seitei jisshi jōkyō: Santō-gun Natsuto-mura no jirei," *Teradomari chōshi kenkyū* 7 (March), 31–49.

———. 1992b. "Warichikō: Meireki san nen no futatsu no shōmon." *Bunsui-machi kyōdoshi* 6, 38–46.

Uchida Ginzo. 1921. *Nihon keizai shi no kenkyū*. Dōbunkan.

Umeki Tetsuto. 1991. "Kinsei nōson no seiritsu." In Ryūkyū Shinpōsha, ed., *Shin Ryūkyū shi. Kinsei-hen* 1. Naha: Ryukyu Shinposha.

Vaporis, Constantine Nomikos. 2008. *Tour of Duty: Samurai, Military Service in Edo, and the Culture of Early Modern Japan*. Honolulu: University of Hawai'i Press.

Wajima-shi Shi Hensan Senmon Iinkai, ed. 1971–1976. *Wajima-shi shi*. Kanazawa: Wajima-shi Yakusho.

Wakabayashi Kisaburō. 1962. "Kaga-han no satogo shinkai." *Nihon rekishi* 166 (March), 2–11.

———. 1970–1972. *Kaga-han nōsei shi no kenkyū. Jō, ge*. Yoshikawa Kobunkan.

Walker, Brett. 2006. *The Conquest of Ainu Lands: Ecology and Culture in Japanese Expansion, 1590–1800*. Berkeley: University of California Press.

Walthall, Anne. 1986. *Social Protest and Popular Culture in Eighteenth Century Japan*. Tucson: University of Arizona Press.

Waters, Neil. 1983. *Japan's Local Pragmatists: The Transition from Bakumatsu to Meiji in the Kawasaki Region*. Cambridge, MA: Council on East Asian Studies, distributed by Harvard University Press.

White, James W. 1995. *Ikki: Social Conflict and Political Protest in Early Modern Japan*. Ithaca: Cornell University Press.

Wigmore, John H. 1967–1986. *Law and Justice in Tokugawa Japan, Being Materials for the History of Japanese Law and Justice under the Tokugawa Shogunate, 1603–1867*. Tokyo: University of Tokyo Press.

Yamamoto Yūzō. 2001. *Kome hyappyō*. Shinchōsha.

Yamamura, Kozo. 1974. *A Study of Samurai Income and Entrepreneurship: Quantitative Analyses of Economic and Social Aspects of the Samurai in Tokugawa and Meiji Japan*. Cambridge, MA: Harvard University Press.

———. 1979. "Pre-industrial Landholding Patterns in Japan and England." In A. M. Craig, ed., *Japan: A Comparative View*, 276–323. Princeton: Princeton University Press.

———. 1981. "Returns on Unification: Economic Growth in Japan, 1550–1650." In John Whitney Hall, Nagahara Keiji, and Kozo Yamamura, eds., *Japan before Tokugawa—Political Consolidation and Economic Growth, 1500–1650*, 339–349. Princeton: Princeton University Press.

Yoshikawa Chōshi Hensan Iinkai, ed. 1993–1996. *Yoshikawa Chōshi*. 6 vols. Yoshikawa-machi, Niigata: Yoshikawa-machi.

# Index

## About the Author

**Philip C. Brown,** who received his Ph.D. from the University of Pennsylvania, is currently professor of Japanese history at The Ohio State University. He is author of *Central Authority and Local Autonomy in the Formation of Early Modern Japan: The Case of Kaga Domain* and numerous articles on Japanese history appearing in the *Journal of Asian Studies, Social Science History, Historical Geography,* the *Journal of Japanese Studies,* and other English- and Japanese-language journals and essay collections.

Production Notes for Brown / *Cultivating Commons*
Cover design by Julie Matsuo-Chun
Interior design and composition by Wanda China with display type in
Post-Mediaeval BQ and text in Janson Text
Printing and binding by The Maple-Vail Book Manufacturing Group
Printed on 60 lb. Text White Opaque II, 426 ppi